VIETNAM
COMBAT MEDIC

VIETNAM
COMBAT MEDIC
A Conscientious Objector In The Central Highlands

Ron Donahey

Deeds Publishing | Atlanta

Copyright © 2018 — Ron Donahey

ALL RIGHTS RESERVED — No part of this book may be reproduced in any form or by any electronic or mechanical means, including information storage and retrieval systems, without permission in writing from the authors, except by a reviewer who may quote brief passages in a review.

Published by Deeds Publishing in Athens, GA
www.deedspublishing.com

Printed in The United States of America

ISBN 978-1-947309-72-2

Books are available in quantity for promotional or premium use. For information, email info@deedspublishing.com.

First Edition, 2019

10 9 8 7 6 5 4 3 2 1

*Greater love has no one than this,
than to lay down one's life for his friends.*
John 15:13

I would like to dedicate this book to SP4 William Allen Gilmore. Allen as we called him gave his life to save a fellow medic on February 21, 1967 in the jungles of Kontum Province west of Pleiku. His name is located on Panel 15E Line 68 on the Vietnam Memorial Wall.

Preface

This book is about life in the 60s seen through one set of eyes. For me it was fun and exciting to experience the new things coming out of the 1940s and 50s. I was born in October 1945 into a family not poor and not middle class. My sister, who was born in January 1943, and I were loved and protected by both parents and we had an extended family living close to us. We grew up in a sheltered home. My mother saw that we attended church every week and she also taught us about God and spiritual things at home. She once told me she would pray for us even before we were born.

When the time came to register for the draft I had easily accepted the idea of not taking another's life. Yet being a soldier was something I also looked forward to being. When I registered for the draft the form had three choices to choose from. One was I would enter the military and learn to use a weapon in combat. The second choice was I would enter the Army and not carry a weapon and the third choice was that I would not enter the military.

My choice was the second one. By selecting this option, I knew I would become a medic, choosing to save lives instead of taking lives. When I chose the second one I did have mixed feelings whether or not to bear arms. I knew that in war people would die and I respected those who chose the first option.

Chapter One

"Where shall I begin," he asked. "Begin at the beginning," the King said, "and stop when you get to the end."
— Lewis Carroll, *Alice in Wonderland*

Suddenly our world of silver moonbeams was transformed into one of moving shadows. The tranquil beauty of the night was shattered as the North Vietnamese began their attack. Enemy tracers etched brilliant green streaks through the blackness, as they seemed to float toward us like glowing baseballs. The calm of the jungle was broken by the concussion of our own artillery and claymore mines. The explosions of grenades punctuated the rattle of machine guns, M-16's and M-79's. Mixed with this confusion could be heard the sound of AK-47's. A sound I would become familiar with before my tour of duty ended.

Shadows cast from trip flares silhouetted enemy soldiers as they moved just inside the tree line. The moment I had thought about for eleven months had arrived. My first experience with combat had begun.

Since the beginning of fall quarter in 1964, I had thought a lot about being a soldier. This was the era of *American Graffiti* and *Happy Days*, but it was also a time of confusion. I found myself at the crossroads of adolescence and manhood. My friends and I were being torn in different directions. Some had already left for college, and others were moving away to bigger and better things, they thought. My world was changing. Having spent more time during my high school years perfecting my athletic skills than my academic abilities, I had little interest in continuing my education.

I made frequent trips to the post office to get brochures telling me why I should join the Marine Corp. instead of the Air Force. "Go Navy and see the World" interested me, but in the end this decision was made for me by an Uncle, I was drafted.

I looked forward to the next couple of years as a time of fun and adventure. Having attended a Seventh-day Adventist high school, I was required during my senior year to take a class called Medical Cadet Corp. This was something like an R.O.T.C. program, but its purpose was to prepare us for entering the military through the draft. It was something like basic training with a focus on being a non-combatant in the Army's medical corp. The course reinforced my conviction of duty to God and country and not wanting to kill another human being.

Our instructors were WWII veterans, and our training taught us to go into the Army as medics and be another Desmond T. Doss. Mr. Doss was a non-combatant medic who, during WWII, earned the Medal of Honor. Sadly, we were to learn that the passing of years brings forth new generations. Mr. Doss was not necessarily a product of military training, but at that critical moment he found something deep within himself to preserve life. His true story can be seen in the movie *Hacksaw Ridge*.

Being in a state of uncertainty, I accepted an invitation from my great-uncle living in Dillingham, Alaska, to come north and experience the 1965 fishing season on Bristol Bay. Having lived in southeast Alaska at the age of five, when it was still a protectorate of the United States, and never forgetting the memories, I quickly accepted.

I tied my boat to the cannery dock one sunny day to pick up supplies, and in my mail was a letter from another uncle. It began, as he begins all his letters, "Dear Ron, GREETINGS."

Some of my fondest memories of youth are my four years spent at Auburn Academy. For most of my elementary grades we lived in the shadows of this school, and during my seventh and eighth grade years I spent much of my free time on campus playing sports with the older kids.

Often, I would arrive at the academy Sunday mornings in time to watch the MCC class. I enjoyed seeing them dressed in their uniforms, marching in step, or at least trying to. The tallest ones would be in front and the others in columns behind them. I would watch and dream of the day when I could be in that class.

The Christian influence I received from my home and church had convinced me on some important issues. One of these was that God did not want me to kill another human being. Though I knew all about the killings that took place in the Old Testament, which I never could quite understand, the idea that we were to love our neighbor as ourselves had formed a strong conviction in my heart. So, it was not out of place for me to agree with the idea of being a conscientious objector. Because of the different meanings associated with the title of Objector, I chose to refer to my choice as a Non-Combatant.

On October 14, 1965, my Mom drove me to the induction center in Seattle. The morning was cool, and it looked like rain. The induction center was located on the waterfront and as we passed the ships tied up at the different piers, I wondered if maybe the Navy might not have been the wiser choice. I kissed Mom goodbye, and climbed the steps of the center, the fear of the unknown heavy on my mind. Inside were other teenagers, like myself, who looked as lost as I felt. It was not long before a corpsman in white ushered us into a cold looking room, and told us to take off all our clothes, except our shorts. For the next few hours we were poked, prodded, and pricked as we followed the painted line on the cold linoleum floors moving from one medical station to the next.

I had just given what seemed like a quart of my blood and was standing in the next line when there was the unmistakable sound of breaking glass. I glanced around the corner and sure enough there on the floor was my blood sample nicely mixed with a dozen others. It was not long before my name was called to have my veins probed once more.

When the medical exam ended, we were taken into a large room. On the stage was a podium with a very official looking seal covering its front, and flags on both sides. On the wall behind was a large portrait of President Johnson. We stood there very quiet, not sure what was coming next.

Soon an officer entered and gave a short patriotic speech, then had us raise our right hand as we swore our allegiance to our country. Once this was accomplished, a tough-looking Captain said he needed some of us to volunteer to spend our two years in the Marines, and a couple of us for the Navy. I could not believe my ears. What an opportunity, and only a moment to decide. Before my mind could react, he stepped to the head of the line and told the first three they were going to the Marines, and the next two that they were to be sailors. That made me first in line. I had been the sixth one.

We were flown to Fort Ord, California and spent the next five days taking tests. These tests are a very important part of entering the military. Those who are already truck drivers are told they qualify to become cooks, and those who are already cooks became truck drivers.

My time at Fort Ord was filled with borderline loneliness, and uncertainty. Everyone I saw was someone new, and for the first time in my life I was surrounded by a totally foreign lifestyle. Most everyone was experiencing the same feelings as I was, and it seemed that the majority felt the need to display the tough guy image. Being vulgar and macho seemed the way to mask their true feelings.

I soon had the opportunity to test a common theory expressed by old soldiers. The theory that says, "Never, never volunteer for anything." My second evening in the Army I was called, along with five others, to step forward. After everyone else was dismissed, the sergeant took us into the office, lined us up, and said he needed a volunteer. I stood watching him and wondered why we were there in the first place. Logic told me we were there for a reason, which none of us knew, so volunteering for the unknown was no riskier than being silent for the unknown. So, I went contrary to all I had ever been told. I volunteered.

The sergeant told me to stay in the office while he drove off with the others. He was back in fifteen minutes and showed me a cot, blankets, and a stack of magazines. While the others walked guard that night, I was to listen for the phone. At 10:00 p.m. or as I was learning 2200, I was told to get some sleep if I wanted to. I woke up at 0500 the next morning, fresh for the new day.

After six days at Fort Ord I flew to San Antonio, Texas to begin basic training. The sun was just breaking the eastern sky when I arrived at the headquarters building and signed the roster. I was the last one on the list for the basic training class 6-B, which would begin Monday, two days away.

Chapter Two

"If all men were just, there would be no need for valor."
Agesilaus, 444—400 B.C.

Over the years I have wondered what my life would have been like if I had not been the last person to join class 6-B. My entire tour of duty would have been different. I would have met other people, and my orders may have sent me to Europe, instead of Vietnam.

Class 6-B was just another training class as far as the Army looked at it. But class 6-B was filled with real people. Guys who were only teenagers, but who would either be men or dead before their military experience was finished. We were one hundred and three raw recruits who were not entirely sure what was supposed to happen.

I had listened to my dad and other old soldiers reminisce about their military experiences. Stories of guys from across the country. I found that things had not changed from their generation to mine. We may have been from different parts of America, and from different ethnic backgrounds, but we all had something in common. We had chosen to be non-combatants and had been raised to honor our God and country.

Our basic training consisted of the regular Army basic, except we had no weapons training. As a rule, the officers and NCO's in charge of us were very considerate of our beliefs. It was the exception, not the rule, to hear rough or abusive language from them. We could leave our personal things on our bunk and be certain it would still be there when we returned. We only had one fight during our training. That happened on the first day of basic, and never lasted much beyond a few words and a roll in the dirt.

I had heard a lot about Army chow, but our mess hall was actually pretty good. We were served a variety of foods with a minimal amount of grease. This, I was to find later, was also the exception and not the rule.

Our Thanksgiving meal was a real treat. We were required to wear our uniforms, which were very bare and green. The only objects on our dress greens at this point were gold buttons, and a black and white nametag. Our officers, on the other hand, were decked out in their finest. Some even wore swords.

There was a Jewish Chaplain who kept walking around the mess hall while we ate. Each time he passed our table he would mispronounce my name and ask if I was getting enough to eat. I had the strong feeling my mother had written him a letter asking to keep an eye out for me.

Being away from home on Thanksgiving Day was a new experience for me but I was not alone so I had yet to go through an acute case of homesickness. So far being away from home was like being at summer camp. Basic training ended in early December and after a two week leave we found ourselves back at Fort Sam Houston to begin our advanced medical training just before Christmas.

We returned to find our training would be interrupted during Christmas and New Year's. We were going to be allowed to leave over Christmas if we so desired. I made the mistake of staying on post. Christmas Eve was the loneliest time I had ever experienced. I was the only one in our barracks and the mess hall that was open was on the other side of the base. I spent seven very quiet days alone.

After the New Year holiday, our medical training began in earnest. Class 6-B had given the Army so little trouble that the decision was made to keep us together for our advanced training. This only helped to strengthen the bond that was developing between us. To our surprise, we were not treated the same as in basic training and were able to take part in more activities on post.

In charge of our advanced training class was a tall Mississippian named, Mark Johnson. Mark had been in the Army reserve for two years while he pursued a degree in Pharmacology, and it was his misfortune to be taking his advanced training along with us. Being a more experienced soldier than we were, plus the fact he was about 6' 6", made him a prime candidate to become our acting platoon sergeant.

Soon after our training began there was an announcement made that was to improve my position on the pecking order a few notches. We were informed that those who wanted to try out for the basketball team could meet in the gym at 1600 hours that afternoon. That was music to my ears.

The only drawback with this announcement was the fact it was made at 0500 in the morning, and if I understood military time that was something like 11 hours of waiting. Maybe not that long but it seemed like it by the time I walked into the gym.

I couldn't believe my eyes when I walked out on the floor a little before the time to begin. About seventy-five other guys were already there, and another twenty showed up just as the try-outs began. Contrary to the military's way of doing things, a second Lieutenant walked into the gym right on time.

We all sat down on the bleachers, and he first asked if anyone had played college varsity basketball. Four raised their hand and were told to go to the other end of the court. Next, he asked who had played four years of varsity high school basketball. Five more raised their hands and were sent along with the college players. At this point I knew I had better get with it or I may never hold one of the coveted twelve spots that were available. When he asked about three years high school varsity I was the only one to raise my hand.

Also trying to earn a spot on the team was Mark Johnson and two other classmates, Larry Kingman and Bill Swanson. For the next ninety minutes we played our hearts out trying to impress Lt. Grinouski. At the end of the tryout, there were still four spots vacant. We gathered around to hear the final decision. After thanking everyone for trying out, Lt. Grinouski read the names of the final four players. They were Mark Johnson, Larry Kingman, Bill Swanson, and me. I couldn't believe my ears. I had made it!

It wasn't long before I began to realize some of the advantages of being on the team. As we were sitting in class there would be a knock on the door and after the teacher would answer it he would turn to the class and call our four names. We were to report to the gym on the double. We never were able to suppress a smile as we gathered our things, and left class to go play basketball for the rest of the afternoon.

One sunny afternoon I was called out of class and told to report to the First Sergeant. When I arrived at his office, Lt. Grinouski was there as well. I was given $10 and a day pass. My assignment was to go into San Antonio on the bus and get a quart of paint the Lieutenant needed.

As I rode into town I was feeling pretty good about this extra pass. Downtown San Antonio was very crowded, and as I stepped off the bus a man came up to me and asked if I had change for a dollar. He held a dollar bill between his thumb and first finger. I counted out the change and placed it in the palm of his hand. As I reached for the bill he turned and was lost in the crowd. It happened so fast I was left wondering what had happened. I started to chase him, but after a dozen steps he was nowhere to be seen.

As the season got underway it became evident that ours was the team to beat. Lamar Thompson was one of our players who had played college varsity. He stood 6' 8" and had the skinniest legs I had ever seen. One day in practice he missed a dunk shot so we were giving him a bad time about his inability to jump. He suggested we put a $5 bill on top of the backboard and then see how high he could jump. This of course we all wanted to see. Once the money was in place he took a couple of running steps and snatched the bill from off the top of the backboard, to the cheers of the rest of the team.

At the end of our training when we received permanent orders, Leg's Lamar was told he would be playing basketball for Fort Sam Houston, while most of the class began their journey towards Southeast Asia. When the season was completed we had swept the league, undefeated.

I never had much playing time during the games, but I considered it an honor to sit on the bench in a reserve role and have the opportunity to practice with the caliber of players I had found myself surrounded by.

Chapter Three

"All religions, arts and sciences are branches of the same tree. All these aspirations are directed toward ennobling man's life, lifting it from the sphere of mere physical existence and leading the individual towards freedom."

—Albert Einstein

Class 6-B was only one of many similar training groups at Ft. Sam Houston. Directly behind our barracks was a Special Forces unit from Ft. Bragg. Special Forces is a highly trained group of professional soldiers, but except for possibly being a little more self-assured, and most of the time bordering on the side of arrogance, they weren't much different than the rest of us.

One of the few duties I had to perform during my training was one night of guard duty. It was my misfortune to draw the 1900 to 2100 shift. It was close to the end of my shift when I rounded the corner and quickly stopped. In the twilight I could see a fight in progress in front of one of the barracks. Not wanting to get involved, I stepped behind a coke machine in hopes it would end quickly. Realizing it was not going to end quickly I blew my whistle and headed toward the fight. When I arrived at the grassy arena the fight had stopped.

The two victors were pacing back and forth in front of the barracks trying to encourage the rest of the onlookers to come on out and play. There was plenty of verbal posturing, yet no one seemed brave enough to even come outside and help the bloodied trooper who was still on all fours trying to get to his feet.

As the winners turned to greet me, I saw the loser crawl through the barracks door. I paused just long enough to ask my two new friends if they were all right and we all three continued on my patrol. They wanted assurance that I would not report their play with the enemy. I felt that under the circumstances their request was not out of line.

The following Saturday night two of my fellow classmates were coming back from a night on the town. Getting off the bus ahead of them was a staggering drunk wearing a green beret. As they walked across the dark parking lot one of my friends slapped the drunk on the back and wished him a good evening. In the process he reached up and took the beret off the guy's head.

Two things happened the next day. In the mail, on its way to California, was a prize trophy, and the other was an ultimatum which came close to putting a contract out on the person who had done such an unthinkable thing. It was obvious that our neighbors suspected our class, but they couldn't prove it.

Chapter Four

"You don't get to choose how you're going to die. Or when. You can only decide how you're going to live. Now."

—Joan Baez

Fort Lewis meant home for me. I would rather have been on my way to Europe, but if not there, home would do just fine. Coming with me were most of my close friends I had made during our training at Fort Sam.

When we left San Antonio, we were given individual tickets, and a set of orders which said we had to report to Ft. Lewis within seven days. On the plane between San Antonio and Los Angeles, I heard others talking about continuing with the group, which would arrive that afternoon, or going home and reporting on the day the orders said. When the plane left Los Angeles there were five empty seats.

My parents were waiting for me at SeaTac airport in Seattle, and my Dad had been talking with the sergeant who was waiting for us. As we began to board the buses for the ride to Fort Lewis, the sergeant told me to go on home, but to report in before the time specified on my orders. This was great news. My old '51 Chevy was due to be replaced, so now I would have time to look for another car.

By the end of the seven days I had found a '63 Chevy Super Sport which I felt would fill my needs quite well. A candy apple red, four-speed, with bucket seats, and a 327 cu. in. motor. When I arrived at Fort Lewis I was riding in style. As I drove into the parking lot I noticed the other five guys just arriving too.

When we reported to the orderly room, we were told we had been AWOL, and would have to wait while someone went to get the Captain. This wasn't the reception we had expected. It appeared we were in trouble on a technicality.

The orders we received at Fort Sam had been group traveling orders, which meant we all were to report in on the same day. The seven days was not to be a part of the orders. Thanks to the unknown sergeant at the airport, they couldn't punish me, and it didn't seem fair to punish the other five for doing what I had done, so we were assigned KP duty the next day. We all felt it was a fair trade for seven extra days leave.

It was March when we arrived at Fort Lewis, and the weather was typical Pacific Northwest for that time of the year. The day we arrived was cold and wet. The barracks we were assigned to were old and used coal for heat. When we were taken to our regular units it was a relief to see new modern barracks, even better than the ones at Ft. Sam.

My new home was Headquarters Company, 2nd Battalion, 8th Infantry Regiment, 4th Infantry Division. Our group, which had been together for the past six months, was now split up in units all over Fort Lewis. We also found out that the 4th Infantry Division was just beginning their advanced infantry training. That meant we were going to have to go through another couple of months pulling medical coverage with the infantry as they trained. This wasn't the bad news though. It was also rumored that once the training was completed we would all be going to Vietnam.

When we arrived at our new barracks, Ruben Martinez and I were assigned a room which was to be for those of the rank of E-5 or above. We knew our privacy was not going to last forever. Soon someone would catch the mistake and we would be put with the other medics in a large bay on the top floor of the headquarters building.

We had enjoyed a week in our private room when our luck ran out. After breakfast we had an hour to clean the barracks before our day began. Our room took just a couple of minutes to straighten up, so we decided to lay under our beds and stay out of sight. Out of sight, out of mind, you know.

We had been there for fifteen minutes, or so, when the door opened, and we saw two pair of spit shined boots standing between our two bunks. One pair belonged to Captain Morrison, and the other to Sergeant Mendosa.

We could see each other under the beds and our eyes were glued to those two pair of boots. Capt. Morrison asked Sgt. Mendosa who was in this room. Sgt. Mendosa replied, "It is Martinez and that other dud." Under different circumstances I would have taken exception to that endorsement, but due to our present location I felt it wise to accept the title.

At the end of their little talk, Sgt. Mendosa was told to put us in with the other medics, so we knew we had slept the last night in our private room. We waited five minutes then climbed out of hiding and headed out of the building.

A week after I arrived I got a phone call from my dad. It seemed that the man I had bought my dream car from was experiencing some marriage problems. His wife had informed him she wanted the car back, and if she didn't get it in a couple of days, she would leave. He wanted to know if I would be willing to return the car. Not wanting to cause them any more problems, I agreed to give it back. To take its place I found a '64 Chevy Nova Super Sport, a four-speed with the same size motor, but all black.

It wasn't long before we slipped into the routine of military life. We no longer enjoyed the country club atmosphere of Ft. Sam, but home was now only thirty-five mile away, and weekends were usually free. I called home and told Mom that I would be bringing a couple of friend's home for the week-end. Word quickly spread that the place to go on week-ends was to Ron's place. After a couple of weeks, I was bringing close to twenty guys home each weekend.

My Mom was on the staff at Auburn Academy and their home was on campus. Being so close to the school, plus Seattle or Tacoma, the guys had a lot to keep them busy. We became a fixture around the academy during the last half of the school year.

It was obvious that my folks enjoyed having such a large group around. My sister and my Mom would cook to their hearts content, and Dad never lacked for help around the house. One of the week-end chores was doing our laundry. One Sunday afternoon my sister washed and pressed twenty-five sets of fatigues. Our presence also had a soothing effect, knowing that in a couple of months we would all be gone. Some may never return whole in mind, body, or spirit.

Our duty at Ft. Lewis primarily consisted of going along with the infantry as they went through their training. We might have to take a jeep and sit at a firing range all day or work sick call each morning. We quickly learned the fine points of being a good medic. Little things that Ft. Sam had forgotten to teach us.

One of these was the introduction to the Army's standard cough medicine called, GI Gin. Appropriately named for its alcohol content. One day I was at a firing range with Art Collins when another of these forgotten lessons was learned.

Things were rather slow, and Art had been nipping at a bottle of GI Gin for some time. Suddenly there was a lot of spitting and sputtering. I looked at Art and saw bubbles foaming out of his mouth. Without realizing it he had taken a nip on the wrong bottle. We learned that when you mix Hydrogen Peroxide with saliva, little bubbles are created and foam out of control. Just another one of the unforgettable moments in our lives. Before our tour of duty ended, Art would experience another, more unforgettable moment in his life.

As the infantry training ended, the climax was to be two weeks out in the field playing war. I was attached to the recon platoon for this exercise. We were to patrol in front of the company and make sure the enemy aggressors were not around.

One morning as we stopped for a break, Lt. Campolo asked me the one question every non-combatant has been asked at least once during their tour, and that is the 'What If' question. He came over and sat down beside me and said, "Doc, when we get to Vietnam, and are out on patrol, and we get pinned down, and I am out in front of you and can't move, would you pick up a weapon to save me?"

Without hesitating I replied, "If you get in that position, and there was a weapon close by I would fire over their heads to keep them down, so I wouldn't have to go get you when they shoot you." He laughed, and said he wanted me as his medic in Vietnam.

As we saddled up to move out across this large open field he was still smiling about our little talk. We made it across without incident and had just entered the tree line ready to radio the main body to proceed across the field, when the quiet was broken by the sound of sniper fire. I ran towards the trees across the road and noticed a group of umpires standing there smiling. We had walked into an ambush.

I was peeking around a tree when suddenly firing opened behind me. I whirled around and looked into the foot-long muzzle flash of an M-60 machine gun hidden in the shadows. That fixed an impression in my mind I would not forget.

Our stay in the States was ending. I was still convinced regarding the idea of not killing another person. Satan knew he couldn't get me to change, so he began to work on the idea of snakes. I do not care for snakes, no matter how small they are. One day we were told that in Vietnam there would be one hundred kinds of snakes. Ninety-nine were deadly poisonous, and the other one would squeeze you to death. This did little to comfort me.

I had begun to consider purchasing a .22 caliber pistol to take with me just to kill the snakes. One afternoon I was heading into Tacoma with Julian White, a medic I had met at Ft. Sam and who was assigned to a sister unit, 1st Battalion, 22nd Infantry, and mentioned this to him. His response was, "What do you think your Mom would say?" That was a good question.

We stopped in Tacoma at a gun dealer and went inside. I found one I liked so I paid the $129.98 and we headed to my folks. I laid it on the kitchen table and looked at my Mom. I saw I needed to be delicate in my response to her inquiry. My Mom would never tell me to not take it with me, but I don't think her fear of snakes was as great as mine.

After spending a few minutes with Mom, my Dad, Julian, and I went to the garbage dump to shoot at bottles and cans. I lined some up and began to fire away. I had shot less than a dozen shells when there was an explosion, and my arm flew past my head. When I looked at my hand, all I was holding was the pistol grip, and part of the trigger mechanism lay across my finger. The barrel and chamber had disappeared, and I had not been hurt. I looked at Julian and Dad, and said, "I think God has just shown me I don't need to be concerned about snakes."

Before we left the States, we were given two weeks leave. Ruben asked me what he could do for me, being as he had used our home on most week-ends. I had already seen a picture of his sister, so I told him I had never seen Disneyland, and would like to spend a couple of days with him in California. So, it was settled, sunny California it was.

When I arrived, and met his family, it was his sister I noticed most of all. Debbe had just turned eighteen, and I realized right away that my stay in southern California was going to be a good one.

The next day I was able to go on a date with her and the guy she was dating. It was July 4th and we went to the fireworks in Downey, California. The next evening, we borrowed Ruben's Corvette, and headed into Hollywood to see the movie, "What did you do during the war, Daddy", at Grauman's Chinese Theater. I had never experienced anything like that evening in my life. Coming out of Hollywood at 2:00 a.m., driving a Corvette with the top down, and it was still 72 degrees.

My week in California was soon past, and while I was there the airlines went on strike. I headed back to Washington on the train. It had taken me three hours to get to California, and thirty-six hours to get back. I stood on the platform in L.A. and kissed Debbe good-bye. When I asked her to write, she said she would.

As our time at Ft. Lewis became short, life insurance salesmen began coming into the barracks each evening. The line at their desk would begin to form early and would still be there late into the evening. The other popular place on post was the EM Club. They sold gallons of watered down beer to guys who were beginning to feel the strain of knowing they were leaving soon. Fights were common. A psychologist would have had a field day observing the behavior of the 4th Infantry Division as time ran out.

The next Friday evening, after our return to Ft. Lewis, our group was preparing to leave for the week-end when a runner from the orderly room arrived and told us Capt. Morrison wanted to see us before we left. We were to go to the Day Room on the double. It was 1900 hours, and two hours later he arrived to tell us we could go on pass until 1400 Saturday afternoon. We were to be back at that time to begin processing for Vietnam. He left saying that it did not matter who we knew, even the President couldn't stop us from being there on our Sabbath.

We sat around after he left and discussed this new problem. It wasn't the idea of being told we were going to do something on Sabbath that could be done during the next week, as much as it was the attitude in which it was given. The consensus was, we felt, it could be done during the week. Saturday was just the first day of seven in which the processing was going to take place.

We decided to call the phone number the church had given us when we were drafted. This was to be used in case we found ourselves in a position which conflicted with our conscience. I walked across the street and phoned Clark Smith, the contact my church had given me. His wife sleepily answered the phone, and told me he was at Camp Doss, and gave me the number. When the orderly answered, I was told to wait as he went to get Lt. Col. Smith. I explained our problem and was told that if we honestly felt it was against our conscience, he would see what he could do. We had been ordered to be at the Captain's office the next day at 1400, so we had to be there.

At the appointed time we were all lined up outside Captain Morrison's door. He walked out all smiles. He looked us over and asked who the ranking man was. Being as three of us had been promoted to PFC recently, and my name began with the letter "D", I was the one. He walked over and put his hand on my shoulder and said he did not know who we knew in Washington, D.C., but we could have the rest of the week-end off. We just needed to be back on post by 0800 Monday morning.

It seems that Clark Smith telephoned his contact in Washington, D.C., who then made a call to the Commander of Ft. Lewis, who began the calls down through the Chain of Command. By about 0400, our Captain had been awakened from a pleasant sleep and asked what he thought he was trying to do. This all seemed rather humorous at the time, but the last laugh was to be on us, once we arrived in Vietnam.

During the last two weeks at Ft. Lewis, we were restricted to base. A couple of days before our departure my family came to say good-bye, and to take my car home. It was an uncomfortable feeling. All the guys who had come home on weekends with me were there saying good-bye. One by one they talked to my folks and tried to give them money as a small token of thanks for their hospitality. Then we were alone. Just me and my family. We had never been a close family where hugs came naturally. Looking back on that evening I know it must have been hard on all of us, not knowing what the next twelve months might bring. After a few awkward minutes, they were gone.

The next Saturday morning we loaded busses that would take us to the waterfront in Tacoma, to board the USNS Pope which would carry us on our voyage into the unknown. While we waited for the ship to leave, the MP's opened the gate and let the family and friends come stand next to the ship. In the crowd I saw my sister and two other girls I had gone to school with. They had seen me and were shouting and waving to get my attention. It was a good feeling to see them one last time.

Chapter Five

"No man is good enough to govern another without that other's consent."

—Abraham Lincoln

Early in the afternoon, the ship moved away from the dock. Soon Tacoma was left behind as we slowly slipped through the waters of Puget Sound. I stood on deck and watched the parts of the world I knew best pass in review. Soon we rounded the tip of Vashon Island, and the Seattle skyline came into view. I looked at the Space Needle and remembered sitting in its shadow on warm summer evenings with my friends at the open-air concerts, and wondered if I would ever see this beautiful city again. Our troop ship dwarfed the Washington State Ferries and pleasure boats that crisscrossed our path.

Once Seattle was lost to view I went below to get familiar with my new home. It consisted of a canvas bunk two inches off the deck, with another one just like it twenty-four inches above mine. Things were in a state of confusion, so I slipped out and went back on deck. We had rounded the point and were passing Sequim Bay. It wasn't long before Victoria, B.C., came into view off the starboard side of the ship and we were soon in the open Pacific.

It wasn't long before we slipped over the horizon. The ocean was calm and would remain that way for the eighteen days we were to be sailing on it. It was a pleasant eighteen days. Full of sunshine, lazy afternoons playing hearts and gin rummy. Each day there would be a movie down in the depths of the ship. Often, I would stand in line for two hours to assure a spot in the theater, just to get there and find out the movie was one I didn't care to see, or to stand in the PX line for a couple of hours and get up to the door just when they closed it for the day.

Without a doubt, the worse part of the voyage was the food we were served. The officer's food was a different story altogether. One day I went below and found a small group getting ready to devour a couple of freshly roasted chickens. It seemed one of them had walked past the officer's mess and noticed what was on their supper menu. He rolled up his sleeves and walked into the kitchen as if he were on KP duty and liberated a couple of chickens that had just come out of the oven. On a bunk next to us was one of our NCO's who wasn't all that popular. He was deep in sleep, and his hand was lying on the deck, palm up. We couldn't resist the temptation of leaving the bones in and around his hand when we left.

Thus inspired, another brave comrade rose, rolled up his sleeves, and told us to be patient, he would be right back. In ten minutes, he returned with a case of oranges over his shoulder. I was glad to have such brave friends.

Two days later, I found myself on KP duty. I ended up in the bottom of the ship with three other guys working the cold storage section. What a pleasant day. Whenever an order was sent below we would gather the food together and up the elevator it would go to the kitchen. It wasn't long before we located the officer's cold storage units, and what an eye opener. Not only was there fresh fruit, but milk as well. That was by far the best day I spent on the cruise.

We quickly learned that the daily amount of allocated hot water never lasted much beyond sun-up. Those who wanted a fresh water shower would be the first to rise each day. Cold salt water showers were never a problem.

When the ship pulled away from the dock in Tacoma, a dice game began in one of the large shower stalls. That game lasted for eighteen days, twenty-four hours a day. It was not unusual to see a pile of money in the middle of the floor.

On the bow of the ship there is a place where the mooring lines go through the side when the ship is docked. These made perfect saddles to sit on and watch the flying fish as the bow cut through the water. It wasn't just the flying fish that was fascinating, we also saw shark, rays, and many other sea creatures. I found it interesting to see sea gulls flying in the middle of nowhere, no land in sight.

I was on deck one day and was lucky to be one of the few who saw a water tornado. One of the crew told me they were called a Sea Spout. He had been at sea for many years and this was the first he had seen.

The next day as I stood at the railing looking out over what I thought was the peaceful Pacific Ocean, I noticed a very small boat off our starboard side that appeared to be a rowboat. On its bow were a pole and a lantern, and in the back was a man busy fishing. The boat was powered by the oldest outboard motor I had ever seen. Little did I realize that we had left the Pacific Ocean far behind and were now in the East China Sea. Okinawa was just over the horizon and we were scheduled to stop there to refuel.

Our stay was just long enough to take on fuel. I stood at the railing looking at the dock just a couple of feet away, wishing we could get off for a while. We did receive mail though, and I received a couple of letters from home. We were more than willing to call our letters a fair trade for having to stay on board.

Later that day we left Okinawa and I realized the next landfall was going to be Vietnam. The upper deck was off limits after sunset, but the last couple of nights on board I went up on deck anyway. I would find a dark corner to hide in and watched the tropical lightening off to the west. At first, I thought it was bombs exploding, just like I had seen on television war movies. As I was hiding in the shadows I would notice others sneaking out on deck, too. They would look in every direction, and then quickly find their shadow to hide in. My thoughts were never about being hurt or killed. Since the gun exploded in my hand just prior to leaving the states, I was at peace about my safety.

The last night aboard ship finally came. We had sailed on three oceans during our journey, and now that it was ending we were apprehensive about what lay ahead. With us were some older NCO's who had fought in WWII and Korea. They too wore the mask of the unknown. We even had a couple with us who had sailed to war in the early 1940's on this very same ship.

Wake-up that last morning was early, 0330 in fact. By sun-up we were all on deck with our gear. As we entered the harbor of Qui Nhon, we noticed a change. Since we had sailed from Tacoma we had felt a cooling sea breeze. Pleasant would describe it best. As the ship slowed to harbor speed, and finally dropped anchor, it began to feel like we were in a steam bath wrapped in heavy clothing. The pleasant sea breeze had vanished and in its place was an odor of stale air.

For the first time in my life, I learned what humidity meant. Even though it was early morning, and the sun had just risen, it was already hot. I stood on deck looking out over the harbor and across the horizon sailed a fleet of war ships. I counted at least five, and one looked like an aircraft carrier.

Finally, it was time for my unit to disembark. The ship had dropped anchor about a half mile from shore and our ride to the beach was going to be in a Navy landing craft. I was the first one to board the craft and was instructed to proceed to the far corner and put my duffle bag on the deck between my feet.

The next person was to position himself against my bag and put his bag on deck between his feet. This was done by everyone boarding until the craft was filled and we were packed in like sardines. Within moments of boarding the landing craft, our fatigues were soaked with sweat. The side of the craft was about chin high, and what little breeze there was on the water that morning was hitting the side of the craft and not going any further. By the time we finally pulled away from the ship, I was ready to pass out from the heat.

As the craft touched the sand, the gate dropped and we surged ashore. It reminded me of the WWII movies of the Marines landing on Iwo Jima. The relief of being out of the landing craft and on the beach, was not much better. The sun glaring off the sand reflected the heat just as much. It wasn't the heat as much as it was the humidity. I had never experienced anything like it before.

We were taken by bus to the airfield and, as always, we waited. The next part of the journey was to be on C-130's for a flight to Pleiku, in the central highlands. By early afternoon we were marched to the waiting planes. Instead of standing shoulder to shoulder, as it had been on the landing craft, we were told to sit with our duffle bags between our legs, and the next person sat with his back against our bag, and so on. When we were loaded there was a solid mass of bodies intertwined on the deck of the plane. By the time we arrived in Pleiku, it felt like we had been through the war and were ready to go home.

We had not been able to see much of the country since our arrival. This was about to change and our eyes were about to be opened to the ways of a country nothing like ours. Trucks were waiting as we landed in Pleiku. Our destination was to be our new home a few miles the other side of the city. As our convoy left the airport and began to wind through the streets of Pleiku City, we were captivated by what we saw. As we neared the main highway we came to a stop. Coming up beside us was an old mamma san wearing black pajama's and a conical hat, carrying two baskets on a pole. She stopped beside us and set her baskets down, dropped her pajamas and proceeded to go to the bathroom not five feet away. She squatted there beside the truck, looking up at us with her toothless beetle nut grin.

Soon we were out of the city limits and heading into the countryside. The reaction to what we were seeing was varied. It reminded me of the pictures missionaries would show at church when they visited. To some the only thing they could talk about was the fact we were driving out into the country without ammo for the weapons they carried.

The country was relatively flat and open, and we soon arrived at a muddy road heading south. The monsoon, with its heavy rain, had turned the hard-packed road into a sea of gumbo. We were told to walk the rest of the way because the road had become impassible. As we walked we could see trucks buried up to their axles. We walked around a hill and entered a large open field. This was our new home. This large empty field was to become Camp Enari, home of the 4th Infantry Division. It would resemble a small town with swimming pool and paved roads.

When we entered the field, there was nothing there but a few tents used by those in the advanced party who had arrived earlier that month. Our first objective was to put up our pup tents. No one was sure when our equipment was to arrive by truck from Qui Nhon, so our two-man tents were to be our shelter. We learned quickly what it is like to live without the bare necessities we were all used to.

I soon found the only way to keep the mud, which was more like red clay, off my boots was to tie sand bags over them. The rainy season in South Vietnam extends from May through November and it rains every day, and when it is not raining the boiling hot sun was a sign it was about to rain some more.

Our time was spent filling sandbags and building bunkers on the perimeter. Occasionally we would watch with interest as Montagnard tribesmen would walk past. Montagnard's are the hill people of Vietnam, and occasionally they would stop, and with interest watch us as we worked. I had never been around people whose clothing consisted of a loin cloth, and a blanket like shawl all woven with the same coloring and design. Few of these people wore shoes, and their belongings resembled something found in the stone age.

Family and friends standing behind the 4th Division Band just prior to leaving

My sister and two friends

Waiting to board USNS Pope at Port of Tacoma

The Domain of the Golden Dragon is an unofficial United States Navy certificate. It was given to us when the ship crossed the International Date Line.

Chapter Six

"Never give in—never, never, never, never, in nothing great or small, large or petty, never give in except to convictions of honor and good sense. Never yield to force; never yield to the apparently overwhelming might of the enemy."
—Winston Churchill

Having lived in the Pacific Northwest for most of my twenty years I was familiar with rain, but not this kind of rain. By the second day, our small camp was beginning to take shape. Our row of pup tents was lined up with basic training precision and we had even dug trenches around them just like we had learned at Ft. Sam eight months earlier.

I had spent the last hour of sunlight blowing up my air mattress, but John had decided it wasn't worth the effort and went to slept on the ground. During the early morning it rained so hard that I floated half out of the tent. When morning came John wasn't in a very good mood. He had spent the night, mostly awake, curled up in a soggy sleeping bag.

The first Saturday morning we awoke to a slight drizzle, and the coolest it had been yet. The humidity and heat of Qui Nhon was only a memory as we milled around in our jackets trying to get warm.

Sergeant Mendosa was a lifer who had already spent over twenty years in the Army. His aspiration was to maintain a peaceful coexistence, and never muddy the water, so he could retire a happy man in just a couple of more years. He didn't look happy as he walked over to us and ask if we would please be willing to work. Saturdays had always been our day off and we were released from regular duties.

It was apparent that Sgt. Mendosa's request for us to work was coming from higher up than his own personal desire to keep us busy. When he couldn't get any of us to agree to work on our Sabbath, he told us we would have to come with him, the Captain wanted to see us.

The sun was beginning to come out and the clouds were disappearing as we lined up in front of Capt. Morrison. The look on his face said it was payback time for the embarrassment of that Saturday morning at Ft. Lewis.

He told us that he had been in contact with our priest who, he said, was in Saigon and we were to be advised that our church was waiving the regulation restricting us from working on Saturdays. He then asked how we felt about the good news.

It was clear to us he had not contacted a representative of our church regarding this new policy change on our behalf. It had always and would continue to be a personal choice of each individual according to their convictions. When it was clear we still felt we wanted our religious freedoms, we were told to sit at attention and reconsider our decision. By this time the sun was out in full force. We were not allowed to take our jacket or steel helmet off, so we sat there bathed in sweat.

After half an hour, Capt. Morrison returned. It would be safe to say his attitude was not pleasant. He ordered us to stand and come to attention. When we had arrived at his tent I chose to be as far away from the sparks as possible. It had worked so far, but it was about to change.

He walked over and stood right in front of me and gave me a direct order to go fill sand bags. As courteous as possible I told him it would be going against my personal religious beliefs. He stepped in front of Clarence Lipscomb and repeated the order and received the same answer. Next, he went to Allen Gilmore and he too refused the order on the grounds of religious freedom.

Frustrated would be describing it mildly, livid would be closer to it. He restricted the other six to their tents and the bathrooms and marched the three of us to have a talk with Lieutenant Colonel (Lt. Col.) Kincade, the battalion commander (not his real name). As we entered the command tent, a patrol had just called in reporting possible sniper fire. This was a platoon size patrol and its range was two thousand meters away from our Base Camp. Being one of the first patrols we had sent out, the pressure of a potential problem had created a tense atmosphere. Lt. Col. Kincade was having a difficult time deciding what to tell the platoon leader. Seeing the embarrassment, a Major quickly ushered us into the next tent and told us to wait.

In about fifteen minutes a very unhappy Lt. Col. Kincade came in and wanted to know what the problem was. Capt. Morrison proceeded to explain what uncooperative young soldiers we three were. When he told the Lt. Colonel, we had refused a direct order that was all it took. I had been standing at attention with my eyes straight ahead. Lt. Col. Kincade jumped in front of me and with his face inches away he told me what a religious jerk I was. With spit hitting my face, and language too graphic to print, he told me that he was twice the Christian I would ever become and told me if I did not obey the Captain I would be subjected to a court marshal and dishonorable discharge.

The entire time he was yelling at me, my eyes were glued to those silver circles pinned to his shirt. From one side I was being told that if I did not obey I would really be in deep trouble, and from the other side I was hearing that soft voice saying that if I would only choose to obey my convictions everything would turn out fine.

When he gave me a direct order to obey whatever the Captain said, I unpretentiously told him I could not obey any order which went against my religious convictions, which was a freedom given me by the Constitution of the United States.

He stood there staring into my face, speechless. And the peace which only God can give filled me completely. I did not dislike this man, and I was not afraid of him, or his power. It was obvious he had met his match. Jesus Christ. He turned to the Captain and yelling told him to get us out of his sight. We were returned to our company area and restricted, along with the others, to our tent and the bathroom.

This whole incident had been witnessed by those in our battalion. Throughout the day some of them who had watched came by and wanted to know what it was all about. That evening Clarence, Allen, and I were informed we were going on guard duty as punishment while those in charge tried to figure out what they were going to do with us three trouble makers. As evening twilight turned into darkness, we were put in a foxhole next to a machine gun position. The next ten hours or so was a real experience.

At this point in our military career, none of us had much experience with guard duty. My only encounter was the one shift during medical training at Ft. Sam. We quickly learned all about fear. More than once we could clearly see the enemy crawling towards us. We were certain they were moving back and forth in front of us.

What we were experiencing was the power of the mind. There was no one out in front of us. Just the fear of the darkness and not being able to see. Occasionally a flare would pop above us, so we could see there was no one out there.

When the sun came up we were sent back to our tents. Throughout the night I had been thinking a lot about what had happened. I recognized the peace I felt inside, and its source. Prior to leaving Ft. Lewis I had been assigned the job of being the medical clerk and assistant to Dr. Lang. A cushy job, safe and away from combat. It just so happened that our medical tents were to be located across the company square from the Lt. Colonel. That idea did not make me very excited. I envisioned always getting KP or latrine duty, and never achieving a rank beyond PFC.

During the night I had decided to request a transfer to a line company. This meant I would be subject to combat, but along with the peace I felt, and never having been in combat, I had no fear of the possible outcome of my decision.

As soon as Dr. Lang was up and about I went to see him. I explained my feelings and logic behind my decision. He was very happy to approve my request. It was also a way to eliminate the pressure he was feeling from the Lt. Colonel.

He took me to the command tent of Alpha Company where he introduced me to a Capt. Summers, as their newest medic. Capt. Summers shook my hand and welcomed me. He in turn introduced me to my new platoon leader Lt. Underwood. I had been around the Lieutenant several times while we were still training at Ft. Lewis. As we turned to leave, Capt. Summers told the Lieutenant that if I ever forgot what day Saturday was, he was to remind me. I would not have any problems with my religious convictions in his company. My new duty was to be the medic for weapons platoon. On our way to the platoon area, Lt. Underwood told me a little about the men I would be with. He also asked what the problem with Capt. Morrison and Lt. Col. Kincade was all about. I tried to explain my conviction to him the best I knew how.

When I finished, he said he may not agree with me on every point, but that he respected me for standing up for what I believed. He then blew me away by asking if he could call me Doc. I had never had an officer ask me if they could do anything.

Like the Lieutenant, most of Alpha Company had witnessed the events of the day before. I was welcomed by most as some kind of hero. I had stood up to the Lt. Colonel, and in their eyes had won. It wasn't hard to fit in with these new guys. We spent five months together at Ft. Lewis and had seen each other before. I was in the aid station the next day when Larry Harshman, a member of weapons platoon, came in. He asked if anyone could cut hair, and without hesitating I said I could. As we went out to the generator I kept telling myself I could cut hair. I had seen it done many times. It can't be that hard. Besides, this would be a great way to get to know Larry. Half an hour later we finished, and I had done a pretty fair job, and I had made a good friend. Larry was married and had a young son.

As I got to know the others in my platoon I learned what motivated them. Most were just like me, spending two years out of our lives because, as we were told, our friends and neighbors at home had requested that we be drafted. We were marking time, waiting until the last mark on our short-timers' calendar was noted, and we were on that plane ride back to the world. Returning to a life which was never to be as perfect as we imagined it would be.

We walked the trails and picked our way through the jungles and swamps of Vietnam and we talked about our girlfriends, or thought we had, and the cars we had or would get once we returned home. The first couple of weeks in Vietnam were spent setting up camp. The only ones who were able to leave were those who were driving trucks, either hauling our equipment in, or carrying garbage out. I was able to hop a ride with a garbage truck one afternoon, and for the first time since we had arrived I was able to see some of the local people.

The truck hauled a dozen garbage cans filled with the slop we scraped from our trays after we had eaten. I had never seen people as poor as those we saw at the garbage dump. I was an American, I had been born in the United States where we had the luxury, if we chose, of throwing away the food we did not eat.

As we entered the gate to the dump I saw people running towards us. They were young children and old mamma sans with beetle nut smiles. Before we came to a stop, those who were fast enough were already on the truck with their arms, up to their arm pits, fishing around in the slop for food they could eat. Fights broke out between them for this privilege.

It was every person for themselves, with no respect of age or gender. As they would find a chunk of meat, they would pop it in their mouth and wolf it down. When they had finished we dumped the remains over the side of the truck where a large pig picked through what was left. As we drove away I looked back and saw an old mamma san fighting with the pig for a scrap of food they both wanted.

At Dragon Mountain garbage dump

Chapter Seven

"A young man who does not have what it takes to perform military service is not likely to have what it takes to make a living."

—John F. Kennedy

The rumor mill was running at full speed by this time. One day we were going to the Delta Region, below Saigon. The next had us headed to a coastal town, called Tuy Hoa, to guard the Vietnamese as they harvested their rice. It seemed that taxes were not just an American phenomenon. The Viet Cong would take part of the rice crop for their war effort and call it taxes. The VC tax collectors had a more permanent method of collecting from unwilling tax payers.

The rumor that seemed to always be around was putting us on the Cambodian border west of Base Camp. We were hearing talk about the 1st Air Cav and a search and destroy operation called Paul Revere. Anything sounded better to us than staying in Base Camp filling sand bags.

Dragon Mountain Base Camp was beginning to look more and more like a military base. We were building bunkers and towers around the perimeter. Beyond these were rows of concertina wire, and fifty-five-gallon drums of a flammable liquid called foo-gas. This is a mixture of jellied gasoline and is to be exploded during an attack. When exploded it spread flaming napalm on anyone unlucky enough to be close.

So far, we had not seen any snakes. The part of Vietnam we were in was quite open, with small trees scattered about. The only animals to be seen were an occasional dog. I had stopped looking for anything crawling and had begun walking with my eyes focused above two feet in front of me.

Things were starting to look up when someone reminded me about the one hundred species of snakes in Vietnam. You know the ninety-nine that are deadly poisonous, and the other one that would squeeze you to death, so it was only a matter of time. That's all I needed. I again went back to looking at the ground. It is probably good that I did, because I would have missed the opportunity to see what a Bamboo Pit Viper looked like.

I was walking down a tank track one day, with my senior aid man John Gillette, and came face to face with one. The fact that it had been run over by the tank did little to comfort us. I'm sure we set world class records for the triple jump. Looking down, as I sailed over it, I noticed a red stripe on its tail. We finally got our breathing under control and decided we just had to get a closer look. It was around three feet long, had a diamond shaped head, and was a very bright green, except for the short red stripe on the tail. After viewing this creature up close and personal we began to see them everywhere. Before my tour was over I saw several dozen of them. They were very common in the central highlands of Vietnam.

I was beginning to believe that anything not green was made of red dust. It was everywhere, and as soon as it would rain it turned into red mud. The sun would come out and it turned back into red dust.

For the first couple of weeks we were left with our own imaginations on how to keep clean. The usual way was to strip and bathe whenever a monsoon rain came over. This was a suitable method except for one flaw. Sometimes it would not rain long enough for you to get the soap off. The monsoons could come without much warning and stop just as fast.

After one of these quick showers I was standing inside a tent watching several guys trying to wipe the suds off before they dried. I noticed that the tent flap had caught a large amount of water. It seemed hard to believe that this had not been detected before. I got my poncho out and lined a sand bagged foxhole. I used my helmet and bailed the water into my bathtub and I climbed in. When I finished there were several guys waiting in line.

Being the medic for weapons platoon was turning out to be a hard job. My most important duty was handing out the malaria pills each day. What work they would let me do was very light. It was as if they were trying to protect me from something.

My transfer to Alpha Company had in fact relieved the pressure from my run in with Lt. Col. Kincade. Clarence and Allen were still in headquarters company and continued to feel the tension. The next Saturday morning Lt. Bradford, the protestant chaplain, came by the aid station and wanted to know if we would like to have a church service. Chaplain Bradford was a very soft-spoken Baptist preacher. His talk was on convictions. He seemed impressed that we would have such strong beliefs and would stand up to a system that placed the importance of religion somewhere on the priority ladder below basic necessities.

Clarence was on duty in the aid station so had to return to

his work after Chaplain Bradford finished. Just as he walked in the tent a Vietnamese man showed up with a deep laceration on his right knee. Filthy was an understatement. I'm sure it had been several weeks since he had bathed. Dr. Lang looked at it and told Clarence to fix it. His only experience with sutures was the class we had taken at Ft. Sam. Not wanting to show any ignorance, Clarence went at it like it was something he had been doing for years. The best thing going for him was the man couldn't speak English. By the time the wound was clean, and he had begun putting the six stitches in, the man's knee was so deadened he probably wouldn't feel anything for several days.

With so much spare time I began writing letters. Our mail system was still sorting itself out. We would go for days without getting any, then it would all come at once. As soon as we had left Fort Lewis, my Mom had begun writing to all twenty of the guys that had shared our home on week-ends. These were not short letters but small books. I hadn't received a letter from home in over a week, and in one day I had three guys come by and tell me they had each received a fifteen to eighteen-page letter from my Mom. I wasn't proud, I read their letters.

Weapons platoon would fire their mortars each evening for practice. They would walk the rounds to about seventy meters in front of the perimeter bunkers. This was good practice, but it also caused our first casualty. The bunkers were built with small firing ports facing the front. One evening a piece of shrapnel came through one of these ports and struck a guy in the arm. It went through his arm and into the chest, collapsing a lung. After getting him bandaged we took him to the aid station where Dr. Lang could check him over. He was then medevaced to the 18th Surgical Hospital in Pleiku for surgery.

A couple of days later, John and I were standing by my tent talking to one of my guys, when there was a tremendous explosion behind us. I spun around and saw a ball of black smoke with an orange center not two hundred feet away. My first thought was incoming rockets. We had not experienced rockets yet, but what I saw seemed to fit what they might be like.

I dove for a foxhole and landed on top of someone else. Looking up I could see the small ammo dump, and a body lying beside it. I jumped up and started running towards it but was stopped by someone telling me to stay back. I waited a few seconds then ran to the body.

It was then I understood why Dr. Lang had wanted each of us to watch at least one autopsy before we left the States. I stood there looking at a body that could not be identified. One leg was gone at the knee, the other just above the boot. An arm was missing at the elbow, and the head was split open.

I turned to ask for a poncho to cover the body and saw another body lying between where I had been, and where I now stood. In my haste to reach the first body, I had not seen the other. I ran to the second one and looked into the lifeless eyes of Sgt. Larry Bingham from first platoon. He was not quite as torn up, but he too had limbs missing.

This happened just before supper so most everyone was in the battalion area. Weapons platoon was in charge of the ammo dump and my friend Bob Stockton had been on duty. The first body was tall, and I noticed the blond hair. Bob was tall and blond. A quick formation showed three unaccounted for. Bob, Larry, and a sergeant from 1st platoon.

As we began picking up the pieces, Bob came strolling into the company area. He had gone to the battalion ammo dump for supplies and had not been there when the explosion occurred. We now knew who the other body was.

The two sergeants had been drinking and began to argue about how a claymore mine worked. They had gone to the ammo dump, and no one was there, they went in and took a mine out. After connecting the detonator, one of them held the mine and the other set it off.

We were all affected by these meaningless deaths. They showed us that it was not just the enemy we must be cautious of but ourselves as well. It was hard to believe that anyone could be so foolish as to hold a claymore while someone else set it off.

Without realizing it, an emotional escape mechanism was beginning to take control. This was part of our psychological make up which gave us resilience to the violence we were finding ourselves surrounded by. We knew we were getting close to leaving camp when the 1st Battalion Twenty-Second Infantry received orders to go to Tuy Hoa for the rice harvest.

Upon their arrival, weapons platoon set up the mortars and began firing. It wasn't long before they were shown the results of their efforts. A Vietnamese farmer came by camp wanting to see a doctor. With him were a couple of his cows that had been wounded by the mortars. Being as it was their fault the cows had been wounded he wanted to know if one of the doctors could fix them.

Their primary job was to clear the VC from the surrounding hills. They began finding equipment and supplies in the caves and they also began making contact regularly. On one patrol, they had stopped for a break and Gene Yost was talking with some of his guys. As he walked away, an enemy grenade landed where he had been standing. His four friends dropped to the ground knowing they would soon hear the explosion and feel the hot shrapnel tear through their bodies. When the grenade did not explode, they jumped up and ran for better cover.

Earlier they had captured a Viet Cong who was hiding under water. It was not uncommon for the enemy to take a hollow reed and use it like a straw to breathe through. Unless you were to look carefully you would usually miss seeing someone under water.

When we had arrived at Fort Lewis, five of our group had been assigned to our sister unit the 1st Battalion 22nd Infantry. Gene had been assigned to Charlie Company and was the first of our group to be wounded. He wrote this in a letter to my folks dated September 1, 1966.

"Today the wind is blowing and it's sorta cold and cloudy. Guess the weatherman decided to give us a few days of cooler weather. Anyway, it's fine with me.

We left Base Camp last Saturday and went to the airport, but my plane didn't leave till Sunday morning. Our operation Base Camp is on the beach close to Tuy Hoa and I stayed there two days before leaving for the first operation. I am in a small village that was hit by the Air Force because the people were helping the V.C. Most of the buildings are blown apart but a few still stand, and I live in one that's C-Companies headquarters. With me are some radio operators the 1st Sergeant and Captain so I feel well protected. Our first day here our mortar put some rounds in a few cows."

In another letter to my folks, dated October 24, 1966, he told about getting wounded.

"By now you've probably heard I am in a different country. I left Vietnam on a stretcher but now am up and around each day here at the 106th General Hospital at Yokohama, Japan. They treat me like a special person here and I am getting better each day.

The morning I was shot will be a morning I will always remember. My platoon was 3 to 4 hundred meters away from 1st platoon and they called for help telling the Captain that 1 was KIA and they had 5 WIA and the medic had shrapnel in both arms. I knew I had to help them. I started around the hill, but a platoon leader wouldn't let me pass so I went back to my position and looked across the long open field that was between me and 1st platoon. The thought came to me to crawl across the wet open field so that's what I began to do. The crawling got tough and soon I was wet all over and my equipment seemed awful heavy.

I stopped for a drink then started again. I kept low and could see I was slowly making progress. I looked up every few minutes just to see I was going straight. I looked up and in front of me was a small green snake. I tried to scare it away, thinking it slither away but to my surprise he lifted his head and just looked at me. I had no idea what kind it was but the vipers we'd been told about were green.

I made a leap around him and kept crawling fast and a minute or two later I felt a sharp pain. I knew I had been shot. The force knocked me on my side and both legs were numb. I laid still trying to think what to do and soon the water around my waist and legs began to turn red. I wiggled out of my gear and reached under my shirt and pants feeling for a wound. Soon I found a small hole in me back, so I felt my side and stomach but didn't find an exit wound. My aid bag was at my feet and knowing I might get shot again I laid very still and put my thumb in the hole. It worked very well and soon the bleeding stopped.

I thank God for saving my life. I laid in that wet field for five hours then crawled to my right to where some armored personal carriers could pick me up. Just before I was pulled into the APC, machine gun bullets again landed beside me going up one side and down the other.

From my ward, I can see Mount Fujiyama if it is clear. It's beautiful to look at with binoculars. Well will close for now. Say hi to Ron.

Sincerely, Gene"

One evening we were given a demonstration of the support that would be available to us once we got to the field. It was very impressive. We saw everything from our own artillery, Phantom jets and Puff the Magic Dragon. Puff was a C-47 loaded with flares and Gatling guns. Each gun could fire fifty-five hundred rounds a minute. The one-million-candlepower parachute flares turned the dark night into brilliant yet shadowy light. The target for the demonstration was a hill close to the camp so we could see the effects of the demonstration.

We received orders the next day that put us on the Cambodian border west of Pleiku City. The 1st Air Cav was putting the finishing touches on operation Paul Revere and we were to carry on follow-up patrols in the border area they would be leaving.

Chapter Eight

"Be convinced that to be happy means to be free and that to be free means to be brave. Therefore, do not take lightly the perils of war."

—Thucydides

For the next two days, we got ready to leave Base Camp to patrol the hills and valleys along the Cambodian border. While we trained at Ft. Lewis we had known it was only a game we were really playing. As the morning of our departure arrived, we awoke with anticipation as well as dread. The game we were embarking on was one that would be played with deadly results. A game decided not in how many points you scored, but in number of enemy bodies.

We were taken to a corner of the camp where choppers could pick us up. Most of us had yet to ride in a chopper. Since our arrival in Vietnam we had watched as they flew overhead on their support missions for the 1st Air Cav.

I looked forward to riding in a Huey but was surprised to see a flight of Chinooks coming to get us. We boarded these graceful machines and took off for the unknown. Civilization was soon left behind. As I stood looking out of the window beside me I notice occasionally a native village pass beneath us. At one point a tea plantation, with its rows of green bushes, came into view. A thirty-minute ride later, we landed in a large field surrounded by the first jungle we had seen since our arrival.

To our surprise, we began building a perimeter. We were the first elements of a fire support base. The artillery was to get there that afternoon. We couldn't win. It was sand bags all over again. Although this time we had a jungle at our door step and bamboo thickets in our backyard.

By evening we resembled a crude Boy Scout camp. The Major brought his command tent and cot while the rest of us built hooch's made from bamboo and a poncho. Five 105mm artillery guns arrived around 1500 and were already firing. Our foxholes looked more like we raked the leaves away and had forgotten to dig.

The jungle is a haunting place. We had seen its vastness as we flew across it and its wildness filled us with a sense of awe. Throughout the day, small patrols ventured inside the jungle to see what was near us; it appeared we were alone. No indication of another human was found. Weapons platoon stayed in camp building positions.

It was now dark. The listening posts were in place, and we sat inside our unfinished perimeter waiting for something to happen. We talked in hushed whispers and watched the trees and brush move. Every once in a while, one of the listening posts would send a breathless report that all seemed to be ok. A comforting message considering no one knew what a dark night in the jungle was supposed to be like.

Around 2100, the Major decided it was time to hit the cot. Everything seemed to be in order. After reminding us about noise and light discipline, he headed to bed. Shortly after he entered the tent a light came on and it sounded like he was involved in hand to hand combat. The language that echoed through the night was pretty graphic.

It never was very clear just what gave him his first indication that everything wasn't like home. When he got his flashlight on he found he wasn't alone. At the foot of his cot was a small family of Bamboo Pit Vipers. He counted four in the family before he was able to get out of the tent. At least now I knew the Major and I had something in common. Our great love for snakes.

Morning finally arrived, and with it word we were going to be leaving. Another company was on its way to guard the artillery. When they arrived, we saddled up and took off into the jungle. None of us knew anything about jungles. We found giant trees, and bushes with large green leaves and vines that wrapped around everything. And on these crawled and slithered the creatures that live in jungles. Creatures that wanted to suck the blood from our veins or sink their fangs into us and fill us with their deadly poison. And we couldn't forget the other one, the one what wanted to squeeze us to death. There was also a third dimension, the enemy who wanted to put a bullet through us.

The Central Highlands was considered by many to be a frightening place. The double and sometimes triple canopy created a gloomy almost eerie landscape that was filled with steep mountains and ragged valleys. The people who lived on these mountains were called Montagnard. Clustered in their villages, they were a primitive and mysterious people. A part of Vietnam's population that reminded us we were no longer back in the good old USA. The Montagnard's are not Vietnamese. They reminded me of aborigines, often living in nakedness and silence. The Vietnamese and Montagnards considered each other the inferior race. Many Montagnards hired out as mercenaries to the Special Forces and proved superior soldiers.

In late 1965, the Ia Drang Valley marked the first large scale appearance of soldiers from North Vietnam. The first AK 47's and RPG 7 rockets were used by the north during this time. The 1st Air Cav won and lost battles here during the year prior to our arrival.

Everything we saw was incredibly new to us. We had been walking for some time when the column stopped. Looking around I saw a three-foot lizard climbing down a tree and we could hear monkeys and birds somewhere in the canopy above us. We had crossed a small stream and had seen paw prints of a large cat, but we still had not seen any human signs.

Every day was the same. Up at sunrise, grab a breakfast from a can, and walk fifteen to twenty klicks before sunset. We began to feel physically good. The walking and climbing were toning us up, but we also began to relax and that could be deadly.

We began to see signs of the enemy, but they were months old. The 1st Air Cav had finished operation Paul Revere on August 25th, after killing 806 NVA. This was now September 6th and we were ready for something more than a stroll in the jungle.

For the past couple of days, we had been dealing with year old punji stakes. Two guys had already been medevaced out because of wounds from these primitive weapons. We found pits filled with these stakes that would have impaled a person if they had fallen into them.

More important than the pain was the infection these punji stakes caused. The enemy had a nasty habit of dipping the tips of these stakes in feces. Infection was a problem with any cut or wound.

We had just finished another cold lunch of C-rations and

moved out when the point man came across new punji stakes. This was the first indication that we were getting close. Another thousand meters and we came upon a deserted village of fifteen huts.

It was an old village in bad need of repair. Some of the huts were ready to collapse. In two of these huts we found cooking fires still glowing. Whoever was there had been in a hurry to leave. Equipment and half empty rice bowls were laying around, but the weapons were gone. It appeared to have been just four or five NVA.

We set up a perimeter and searched the village hut by hut. No one had lived there in a long time. After finding nothing else we began to burn the village down. It was hard to visualize humans living in such conditions.

I was sitting on a log thirty feet or so from a burning hut watching to see what would come out of the thatch roof. I had seen some spiders and a small lizard or two drop out when I looked down. From the hut was a visible line of something gray moving past the end of my boot. It appeared to be lice.

The discovery of this village and the enemy who had been using it was enough to break the lethargic spell we were falling into. We came across an open field around 1700 and decided to stop for the night. Our foxholes were a little deeper and our senses a little sharper as darkness came upon us.

Things seemed out of place. Usually there were jungle noises but tonight things seemed strangely quiet. The silence was becoming intolerable. Suddenly a trip flare went off and a machine gun began firing. I went out one end of our hooch, and the other two went out the other end.

As we lay in the grass straining to see into the darkness, a parachute flare exploded into brilliance, lighting the field and surrounding tree line. The flare made a wavering yellowish glow. The flare burned itself out just as another one exploded over us. The firing had stopped on the perimeter since there was nothing visible to shoot at.

We thought we had seen something just in the trees a little to our right. In the distance, we heard the sound of an artillery gun. Within seconds a round passed overhead and exploded in the trees 150 meters beyond us. I heard the artillery FO ask for another round, but this time drop it 50 meters, and right 50 meters.

Within seconds another round came overhead. This time the sound was louder and faster. When the shell landed, we could see it was right on target. The FO told the artillery battery to fire for effect. Soon five shells exploded, flashing briefly as shrapnel flew through the air. Then another series of five were on the way as the fire support base got the rhythm and then there was silence. Throughout the night, more shots were fired. Chances were that what we had seen was some kind of nocturnal animal instead of some enemy soldiers.

We began seeing more and more signs of the enemy. We had already been on patrols for over two weeks and it appeared we would be there a lot longer. It wasn't often that we got hot food and mail, but we always looked forward to getting both. Our first choice was mail, then hot food.

Word filtered down that we were going to relocate later that morning. Any change sounded good to us and it was going to be done by choppers. As the Lt. Colonel flew around looking for a large LZ, he spotted what appeared to be a full NVA regiment not far from our patrol area. Because of this new sighting, we continued to patrol where we were. At least now we knew there was someone other than us walking around out here.

We began finding fresh signs that appeared to be only hours old. And then just as we began to relax there was a single explosion at the head of the column. A machine gun opened up and then four or five M 16's. Another grenade went off sending shrapnel flying through the jungle. And then there was silence. We had stumbled upon a campsite and in it were weapons, food on the fires, and hammocks still swinging. Whoever had been there had left in a hurry. It had been a small unit but now they would be without weapons and supplies.

We knew the enemy was there but all we were doing was walking day in and day out. The units that had gone to Tuy Hoa to guard the rice harvest had been in combat almost from the time they arrived.

Even though the sun was up and shining brightly, little light filtered through the double and in some places triple canopied ceiling of branches and vines. It might be cooler under the canopy but the humidity, the unseen enemy in Vietnam, seemed to invade everywhere.

At noon we took an hour break and ate another cold meal of C-rations. I was fortunate enough to trade my ham and lima beans for something a little more vegetarian. I put a lot of salt on my food. Some of the reasoning was to replace the salt I was losing because of the heat but the more logical reason was to make it taste more recent than the 1954 canning date stamped on the underside of the cans.

After our cold lunch, we saddled up and took off once more. I began to realize that the real enemy was the sun and humidity. With each step, I could feel the sweat being sucked out of me right through my jungle fatigues, soaking every inch of my body. My wet underwear was riding up my leg and beginning to rub.

The more I walked the more it rubbed, making each step a painful searing agony. I had been warned about wearing underwear in Vietnam now I was beginning to understand.

Around 1400 we were picking our way through an unusually thick jungle. As we passed a ravine on our left someone noticed the top of a hut. We surrounded the ravine and crept closer. It turned out to be a new hut and was filled to the peak of the roof with rice.

After checking it for booby traps, it was decided to blow it up with C 4. While the explosive was being placed, a three-foot cross bow was discovered. This powerful weapon was a shock to our system to find. We had just assumed that if we were going to be shot at we would at least hear it go off. This weapon was silent, and the arrows were razor sharp. Their tips had the appearance of having been dipped in some nasty potion.

By the time we finished, it was close to being dark, so we stayed there for the night. This was the first night we had spent in such dense jungle. The light from the moon and stars could not penetrate the thick canopy so it was darker than we were used to. Even the night noises sounded different.

It was a good feeling when morning arrived, and we had made it through without incident. We awoke with a little more respect for the jungle. The darkness not only protected the enemy but us as well.

As we were leaving we came across a graveyard. The point guy almost died on the spot. He stepped around a giant teak tree and came face to face with a statue guarding the place. After the night we had just spent, that was almost too much for him.

There were only two graves and they appeared to be old as the burial huts were overgrown with vines. This was a perfect place for the enemy to stash supplies, but we decided to let the artillery use it for target practice. We stayed just long enough for the FO to have a marker round put in. This showed us the exact grid location on our map. After we had cleared the area the artillery could fire to their hearts content.

As we saddled up again we noticed what looked like an earthen jar. Taking a closer look, we found not one but five of them. Inside of these jars we found medical supplies from France, China, Russia, and the United States.

Lt. Underwood was holding one of the jars looking at some marking on its side. Out of the jar crawled the biggest ugliest spider I had ever seen. It was larger than my hand and on its hairy back was a bright yellow marking. John Wayne would have been impressed with what happened next. Lt. Underwood dropped the jar and with only one shot from his .45 he blew the spider away. Very impressive. We had seen enough for one day. We saddled up again and headed for more mundane territory.

When we stopped for lunch we were delighted to see a Huey coming in with supplies and mail. This was the first mail we had seen in over a week. I was fortunate to get three from home and one Auburn Globe News. It was nice getting letters, but the paper was like gold.

Along with the supplies we received a ration of beer and pop. The guys knew I never drank beer so they all wanted mine. Give them, I would not. Trade them, you betja. One beer for two cans of pop. I knew I would eventually find someone who thought that was a fair trade. To my surprise no one wanted to take me up on it. I didn't care because I had a plan.

Two days later we stopped for a break after walking close to eight klicks in the heat and humidity. I waited until everyone had sat down and were drinking their tepid Halazone water. I reached into my pack and pulled out a warm beer. Things became deathly silent as I popped the top and sat there looking around at thirty guys about ready to kill me. "Anybody willing to trade their next two cans of pop for this one beer," I asked. Surprisingly they all wanted to trade. From then on whenever we received our weekly ration of beer and pop I ended up with six cans of pop.

It seemed we were just wandering aimlessly, not quite sure where we were or where we were going. We hadn't seen anything in several days that seemed important when all of a sudden, we entered the same village we had burned to the ground a week before. To our surprise, eight of the huts had been rebuilt. As we set up a perimeter we saw three NVA jump up and run. We opened fire on them and hit the front guy, blowing his foot off.

After patching him up, we were able to get a medevac in, so he could be taken to Pleiku for surgery. We also were told we would be going into Base Camp the next day. We were ready. It had been three weeks since we had left. Yet it seemed like a century and a thousand miles ago. We had to travel another ten klicks to a field large enough to get the flight of Hueys in to pick us all up.

The next afternoon we were there waiting for our ride. We soon began to hear the sound of many choppers heading our way. When they came into view we counted over twenty Hueys. Along with these were a flight of A1E Skyraiders and Huey gunships to pull protection for our withdrawal. What a magnificent sight.

The ride back to camp was like no other I had ever been on. All the Hueys traveled with their side doors open and I was lucky enough to sit in a seat for two that faced out the side of the chopper. We flew at over a hundred knots with our boots half out the door. What a feeling. It sure beat walking in the jungle.

Chapter Nine

"Once we have a war there is only one thing to do. It must be won. For defeat brings worse things than any that can ever happen in war."

— Ernest Miller Hemmingway

Base camp hadn't changed much while we were gone except for more towers and bunkers. The mud was just as deep, and the same old tents were waiting for us. Although it was a pleasant change from sleeping on the damp ground. We had been wet most of the time during the past three weeks. It wasn't always the rain that soaked us as we would usually find a river to cross just about sunset.

For the first couple of days, I found myself sleeping a lot. The safety of camp and the added warmth of a tent and sleeping bag allowed my body to relax. I noticed something different in the aid station. During sick call, there was now a separate line just for penicillin shots. At the head of this line, hanging from the tent poles was a syringe about two feet long and a needle just as long. Under this was a sign that said, "Because we care."

Day passes into Pleiku City were available and it wasn't just the trinkets and hangovers the guys were bringing back. Treating sexually transmitted diseases had become the most common sick call function. The standard treatment was 10cc of penicillin each day, for five days.

VD wasn't the only problem we were seeing. Malaria was beginning to raise its ugly head. The symptoms were easy to identify. The high temperature was the easiest symptom to recognize.

I was in the aid station the second night back when a guy walked in, his face contorted with pain. Hanging from the outside of his left hand was a kitchen fork. He had returned late from guard duty and being tired his only desire was to sleep. Without using his flashlight, he began brushing the things off his cot. He had forgotten his mess kit was lying loose and had run the prongs of his fork into his hand just below the little finger. Two of the prongs were embedded completely.

After telling Frank Colburn, or Corky as his friends called him, and me to stand behind the patient Dr. Lang began talking to him as if nothing was wrong. He gave him an intravenous shot of morphine and moments later pulled the fork out. The guy thought morphine was the best thing ever invented. It didn't take long before he passed out and fell backwards into Frank and my arms. We laid him on a stretcher to sleep it off.

Several days later, Frank and I were given a day pass into Pleiku. Corky was also attached to Alpha Company and while in the field we usually were together. As soon as we jumped off the truck we were approached by this kid about eight years old. He turned out to be a pimp for his #1 sister only fourteen and for just $10 he would introduce us to her.

We took off down the street and another group of kids accosted us. This time they began jumping up on us and hanging on. Suddenly, they were gone and so was Corky's wallet with his

ID card and $20. I had started carrying my wallet in my breast pocket with the button fastened so we weren't totally out of money.

Everything we saw in Pleiku City was incredibly new to us. The different kinds of shops, the people, and the language. The traffic was different than I had been used to in the States. There were no modern Chevys or Fords. Sandwiched between the military trucks were old cars that had been painted a hundred different colors. There were Pedi cabs holding one or two people, and Lambretta's which was a scooter like vehicle with a pickup truck bed with a bench seat on both sides to carry passengers.

We spent the day going from shop to shop looking at everything. We bought pure silk for $.80 a square meter and silk bath robes for $4.50. When the sales lady asked me who it was for I told her it was for my mother, so she gave me a mamma san size.

After a while I needed to find a restroom and found that the only ones were in the bars. By that time, I wasn't too concerned about where they were located. I went into the first one I found and walked to the back of the room where I noticed an open door. Sure, enough it was what I was looking for except it resembled an open porch. The lower half of the walls were boards, and the upper part was screen.

As I looked through the screen I noticed it was difference from the street. The front side was a multi-colored mixture of shops and bars. They sold whatever a soldier wanted from brass trinkets and silk to young girls selling the only thing they possessed, themselves. Beyond the screen, I saw the houses of the very poor. Families living in hootches made from discarded ammo boxes. Children playing near open sewers, the air filled with the stench of unwashed bodies and stagnant water.

As I stood at the sheet metal urinal looking out over this tragedy, I detected movement to my right. Assuming it to be just another GI from the bar. A moment later a sweet little voice said, "Hi GI, buy me a drink." I looked at the toilet and saw one of the bar girls sitting there doing her thing and looking sweetly up at me. As I laughed to myself I never thought of the drink she had asked for, but I wondered what my Mom would think if she could see what had just happened.

As I walked back through the bar, I saw the GI's and the bar girls who were trying to get them to go upstairs. Upstairs to the small curtain enclosed space I had seen as I walked to the bathroom in back. Curtains which never quite reached the floor concealed the uncovered filthy mattresses. In a year, the faces would all change, but the game would be the same.

I told Corky what had happened, and we laughed until we had to sit down. Little kids would walk past us and say, "Dinky Dau GI." We walked into a cafe and ordered lunch. He decided on a hamburger and rice and I ordered a steak and potatoes special. For $2.95 I felt I couldn't lose. When our meal came, it looked like any other we had seen. The steak did look well done although I wasn't quite sure about the potatoes. They didn't quite look like potatoes.

I should have asked where the steak came from before I ordered. I could hardly cut it with the knife. One small bite could have lasted ten minutes if we had time. It was quite tasty but the toughest I had ever bitten into. I could only eat half of it. Later I was told it was buffalo meat and the potatoes were some kind of root and were actually very good. The laugh came the next day when the cafe was closed for serving dog meat for their hamburgers.

As we were leaving the cafe, a kid asked if he could shine our boots. Without waiting for an answer, he began to knock the dried mud off Corky's boots and took out his polish. He told us he was fourteen years old and shined shoes, so he could buy food for his mamma san who was sick. I wasn't quite sure about his story, but I was sure about the clothes he wore. The shirt was about ready to fall off and his pants had several large holes in the legs.

When he finished I asked how much we owed him? He said to pay him whatever we thought it was worth. I took him next door to a shop and for $1.50 bought him a new shirt. It was worth it just to see the tears come to his eyes. That he could not hide. He was so excited he just had to have a picture taken with us, but it had to be an even number of people in the picture. Three would not do so he grabbed one of his friends and the picture was taken.

It was time to return to Base Camp and as we walked back to catch our ride a little guy about seven walked up from behind and took my hand and walked with us. After a couple of blocks, I bought him a package of gum and he went on his way.

When we got back to camp we were greeted with a new wood floor and wood frame in our tent. At least now the cots would be level. It really didn't matter much because we also got word we were going out in the field again the next day.

I was told that Dr. Lang was looking for me, so I went to the aid station and found him drinking whiskey and playing poker with two other officers. He just wanted to tell me that he had received good reports and was impressed with my work and attitude while I was out in the field on our last mission. So, because of the power he had, he was making me assistant senior aid man of Alpha Company. I thanked him and told him if he really wanted to impress me he could come up with a promotion to Spec 4. He said he wasn't impressed that much so if I didn't have any money he could win I was dismissed.

This time our destination was the Cambodian border. Thanks to a company from the 25th Infantry Division, a fire support base was already operational. We arrived to replace their company guarding the firebase. As they were getting ready to leave, I asked one of the medics about the place. He said that the official border was about two thousand meters to the west and there hadn't been any action in a long time. Not even a sign of the enemy.

The camp had a battery of 105mm guns and 4.2mm mortars. The jungle was fairly open around us and a stream about waist deep with a sandy bottom flowed through the perimeter. The monsoons were tapering off, allowing the temperature to climb to over a 100* every day.

Just as it was getting dark the first evening, two guys were placing a claymore out in front of their position. While one of them was holding the blasting cap, it exploded, blowing one of his fingers off at the second knuckle. We were able to get him medevaced out before it got dark.

What began as a quiet and peaceful camp turned out to be quite different. The first week we called a medevac every day. The sun was taking its toll, too. More than once I had to put someone in the stream because of heat exhaustion. There were broken ankles and lacerations from machetes.

For the next three weeks, we wandered around looking but finding nothing. Base camp had established a forward camp at a place called Three Tango. This was a large camp and very well protected. It was an old military position that had been used by the French.

We read in the Stars and Stripes that a Buddhist monk was going to Saigon on October 18 to wave a palm branch and blow a bugle. When this happened, the war was to end, and peace would reign forever. On the 18th we were all rooting for him but still planning on the next mission.

The 25th Infantry Division was patrolling a mountainous area twenty-five miles northwest of Three Tango and had been in light combat. What they were finding indicated there might be a large number of NVA there. Our new AO was going to be just west of theirs.

On the morning of the 22nd, we boarded Hueys and headed north. We circled a couple miles away for fifteen minutes as the artillery finished unloading several hundred rounds into the LZ. As we circled, I noticed large craters like giant stepping stones. These were the results of a B-52 strike that had gone in the night before.

Sitting in the cargo bay waiting to descend into the LZ was very uncomfortable. It was nice and cool at fifteen hundred feet, but we knew it was going to be a steam bath once we landed. The day before it had been over 100* and today was no different. Each of us was heavily loaded. Each rifleman carried over 500 rounds for his M 16, three or four hand grenades, several smoke grenades, an entrenching tool, at least two canteens, and usually every other one had a machete. Besides this we were carrying four days' rations. The M-60 machine guns each carried 2,000 rounds and the M-79 grenadiers had between forty and fifty grenades. Each squad carried claymore mines and LAW rockets.

This combat assault was different than the other ones we had made. During the first week of October, a Special Forces cross border patrol had found a newly constructed NVA hospital that could care for over eight hundred patients and was just inside Cambodia. Around the 10th, another patrol had intercepted a large rice carrying party of NVA about three miles northwest of the Special Forces camp at Plei Djereng. That same day over sixty sampans and other water craft had been spotted on the banks of the Se San River. The latest intelligence reports we had estimated maybe two NVA Regiments some place in our new AO.

We all knew it could be our first hot LZ. The choppers came in fast, machine guns blazing, while gunships and A-1E Skyraiders dropped rockets and cluster bombs in the tree line on either side of us. As soon as the chopper landed we were out and running. We ducked beneath the spinning rotors and ran towards the tree line. The choppers hovered in a cloud of grass and dust then took off. As we neared the jungle we stopped and crouched, studying the ground ahead of us.

With the noise of the choppers gone, the silence rang loudly in our senses. A squad broke off and entered the jungle. Soon they were back with the good news that we had landed at an unoccupied LZ. Maybe we were going to be lucky once more. No enemy in sight.

The engineers went right to work with their Bangalore Torpedoes and C-4 and we soon had a clear field of fire around the perimeter. Next the artillery arrived and by night we were looking like a real fire support base. We had been building bunkers and placing concertina wire ever since we arrived. Soon the guns were firing in support of the 25th Infantry Division a few miles away.

We may have landed at a cold LZ, but the jungle around us showed that we were not alone. Three days later, our bunkers resembled fortresses. Throughout the night our artillery had been firing mission after mission. Empty casings were scattered everywhere.

One of the artillery Lieutenants asked our Major's permission to remove a radio antenna from the top of his bunker. His gun was firing just over the top of a nearby hill and he was afraid a shell might hit the antenna. The Major came out just long enough to look and inform the Lieutenant that his gun had plenty of room. Two shells later, a 105mm round hit the antenna and exploded over us.

Three calls for a medic rang out. I ran to the closest wounded and found a guy holding what was left of his right hand. It was gone except for the thumb and part of the little finger. Another had a severe knee injury and the third had been hit on his helmet with shrapnel. It had knocked him silly but had not wounded him.

Mail call on the 26th was good to me. It was my twenty-first birthday and not only had I received eight letters but also a package from my folks with two birthday cakes in it as well. The cakes were incased in popcorn and had arrived in one piece. I kept half of one for myself and set the rest on the bunker for the others.

I set down beside them and began to write a letter home. We were becoming a very busy fire support base. Supply choppers were coming and going on a regular basis. One of the Hueys had dropped its sling load of ammo and was trying to set down beside it. For some reason, the pilot decided to move forward when he was about a foot off the ground. In doing so he caught the front of the skid on a root and got hung up. Instead of hovering there while someone cut the root away he tried to lift the root out of the ground.

He had almost gotten away with it when the chopper lost its power and the Huey dropped about three feet. The impact caused the tail boom to snap just behind the main body. When it hit the ground, I saw the pilot and copilot both slump forward and stay there. By the time I reached the chopper, they had been removed and were just coming to. Both were pretty shook up and one had a nasty cut on his face.

After the pilots were medevaced to Pleiku, I returned to my letter. Sgt. Moran came up and asked if I had been sitting in that same place when the chopper had crashed. I said that I had, and he told me to look behind me. There in the mortar pit buried in the ground was a four-foot piece of rotor blade. The angle it was at and the direction from the chopper had it passing right where I had been sitting. We stood there in silence. Again, I realized the protection I had just received and its Source.

When evening finally came, we were all glad that the day was over. Even though we had seen more senseless accidents, no one had been killed. Bodies had been broken and people's lives altered forever, but we were alive.

Chapter Ten

"If a man were to know the end of this day's business, ere to come, but it suffices that the day will end, and then the end be known. And if we meet again, then we'll smile. And if not, then this parting was well made."

—Julius Caesar

Our intelligence people had hit it big this time. Ever since we had come out into the field on September 5th, we had walked miles and only had a sniper or two fire on us.

On the afternoon of October 28th, Charlie Company came under sniper fire and had one KIA. They were able to confirm three NVA killed before it got dark. During the night one of the three man listening post failed to report in on schedule. They were positioned about a hundred meters outside the perimeter. After they failed to report the next hour, two guys crept out of the perimeter to find out why. When they located the listening post, they found all three men dead.

Bravo Company was also engaged. Throughout the afternoon, they too had come under sniper fire and a constant pressure from a small NVA force that kept nipping at their flanks. They were able to make it to a small hill and had barely begun to dig in when they came under heavy fire. During the night, Bravo Company had four killed and twenty-six wounded.

Jon Dankel, one of the medics, was sharing a foxhole with Sgt. Roy Perlmann, when a mortar round landed on top of them. It decapitated Sgt. Perlmann and severely wounded Jon. He spent the rest of the night laying in the foxhole next to his dead friend.

When it began to get light, the NVA melted back into the jungle, leaving just enough snipers to keep the movement at a minimum. The jungle was so thick that the choppers were having difficulty getting the dead and wounded out.

Our fire support base had been firing continuously since around midnight. Over fifteen hundred rounds had been fired in support of Bravo and Charlie Companies. There was also a group from the 25th Infantry Division being hit as well. They were about two klicks east of the fire base, and had picked up a lot of injured, and had some KIA's. One medevac chopper was shot down, killing all seven on board.

As soon as the sun was up, the Air Force came in full strength. A-1E Skyraiders and Phantom jets were everywhere dropping napalm and five hundred-pound bombs. Huey gunships could be seen circling at a distance, waiting for their turn. Occasionally, a Huey would land inside our perimeter to inspect bullet holes they had received.

Word came down that reinforcements were desperately needed for Bravo Company. They were surrounded and could not move until the NVA broke contact. It was decided that Alpha Company would send its 3rd platoon and one squad from weapons platoon. We were to leave as soon as possible. Third platoon's medic was in Base Camp, so I began to prepare to take his place.

I loaded my rucksack with extra medical supplies and waited along with the others until around 1500 that afternoon. We boarded Hueys and took off. Because of snipers, we had to circle for twenty minutes until it was safe enough to go in. The hill was just large enough for one chopper to come down at a time.

When I landed, I saw my good friend Harold Stenseth crouching in a foxhole. He looked very tired and was just as glad to see me as I was to see him. When I got to his foxhole I began to smell an odor I had never smelled before. A sweet pungent odor that made me nauseated at first. Laying all around the perimeter were NVA bodies already decaying in the heat and humidity of the afternoon.

I couldn't spend much time talking because we only had a couple of hours to prepare our own positions for the night. I began digging with the Lieutenant and Sgt. Moran. We were able to dig a position large enough for all three of us to kneel comfortably and have our bodies below ground. We had brought in lots of sand bags which we used to fortify our hole. I positioned some sandbags and put my aid bag behind them, so I could grab it when I had to go.

As darkness came, everyone seemed prepared. Our position was half way down the hill, and just behind us were my guys from weapons platoon. Fifty feet below us was the perimeter, the first line of defense, and directly in front of us was a machine gun position. The rest of Alpha Company was inserted into the perimeter. The hill was sloped just enough that we could fire over the others without fear of hitting one of our own.

We had decided on a casualty collection point on top of the hill, and although we were still a medic short, we felt we were ready. Harold's position was a hundred feet away and to my left. Throughout the day, we had placed squad size observation posts out in front of us on all sides. Now that darkness had come, the world was ghostly silent around us. The three-man listening post slipped silently out of the perimeter and were soon lost in the blackness. The LP's were to be an advanced warning system in case the enemy began sneaking up on us. The LP's would go out with a radio, their weapons, and a claymore mine or two, and stay there until morning, or until the enemy got too close. If they had to come in during the night they were to explode their claymores, and make sure they brought the radio and their weapons.

The NVA usually did not attack early in the evening, so we were pretty sure it would not happen for a few hours. We went on half alert until midnight, and then full alert from then on. The CP was close to our position, and occasionally, we could hear the radio. The fire support base knew exactly where we were, and every so often the Artillery Forward Observer (FO) would have them fire some rounds around us just in case the enemy was close. It wasn't a matter of hiding from anyone. Charley knew exactly where we were, so it was just a matter of waiting until we got hit.

In the distance, we could hear the muffled sounds of artillery being fired in support of Charlie Company. A full moon was shining under scattered clouds, and our eyes had adjusted to the darkness of the night. It was now over four hours since the sun had gone down, and I doubt if anyone was sleeping. It was very peaceful.

Our listening posts began to report sounds of movement. The brief transmissions had been tinged with panic, a few words, almost frantic description of NVA movement too close for comfort. Suddenly the world of silver moonbeams and soft grays was transformed into one of moving shadows. The tranquil beauty of the night was shattered as the North Vietnamese began their attack. Their tracers etched brilliant green streaks through the blackness, as they seemed to float towards us like glowing baseballs. The calm of the jungle was now broken by the concussions of our own artillery and claymore mines. The rattle of machine guns, and M-16's was punctuated by the explosions of grenades. Shrapnel and bullets slammed into the jungle like the rains of the monsoons. Mixed with this confusion was the sound of AK-47's. A sound I was to become familiar with before my tour was completed.

Shadows cast from trip flares silhouetted enemy soldiers as they moved just inside the tree line. The moment I had thought about for eleven months had arrived. My first experience with combat had begun, and with it came an urgent call from just in front of us for a medic.

As I jumped out of the hole, Lt. Underwood grabbed my arm and said, "Doc, take my .45." I told him I never needed it and took off. As I slid and crawled down to the perimeter, I could see twinkling lights, like fireflies, pointed at us only a hundred feet away. Occasionally an explosion from a Chicom grenade would cause me to duck. Our machine gun was laying down a steady stream of red tracers. At least it wasn't that position that needed help.

As I neared the perimeter, the firing in front of me began to slow and finally came to a stop. Not one bullet was coming in or going out, yet the rest of the perimeter was a continual sound of battle. The fire support base had their rhythm going and was putting a steady flow of artillery around us.

I came to a position and asked who called for a medic. When I realized it was the machine gun position, I thought, "Good choice, Donahey. The one gun the enemy wanted most, and you had to stop here." At that moment, the moon came out from behind a cloud, to reveal in sharp relief a body lying face down out in the field of fire just feet from the tree line.

Without thinking, I ran out to it. When I got to him he whispered to me to stay down. The person who had shot him was in the trees just about 10 feet or so away. I asked him where he was hit, and he said it felt like his hip. I could see a blood stain, so I gave him a shot of morphine and crawled back to the machine gun position. I told the guys I needed help, and one of the guys jumped up and we ran back out. We each grabbed a shoulder and a leg and ran back inside our perimeter.

When we reached the top of the hill, the perimeter where we had just been erupted into fighting once again. As we reached the collection point, Harold came towards me carrying someone else wrapped in a poncho. I asked him if it was one of his guys or one of mine, and he said he thought it was one of mine. I knelt beside the body and recognized my good friend Larry Harshman. He had been hit in the back of the head with shrapnel from a grenade and was in shock. Even in the moonlight, he looked worse than the training films at Ft. Sam.

After getting the two wounded bandaged and as comfortable as possible, I headed back to my position. The firing seemed to be tapering off some. When I reached my hole, I asked the Lieutenant if there had been any more calls for a medic. I was glad to learn that there hadn't been. I continued to the perimeter and crawled into a foxhole with two guys from 3rd platoon. It was then I learned that the guy had been in front of us on a three-man listening post, and that only two of the three had made it back inside the perimeter. The other one was still somewhere outside. I asked if anyone knew who it was, and it was believed to be someone from Bravo Company named Kendel Washington.

The firing had all but stopped by this time. I looked at my watch and realized the battle had raged for almost an hour and it had seemed only a few minutes. It appeared we only had the two WIA's and the MIA. Throughout the night, we could detect movement and noises coming from just inside the tree line. We prayed it was not the enemy grouping for another assault. At this point, all our claymores had been exploded and the trip flares had been used. As the night wore on and we were not attacked again, we realized it was the NVA taking care of their dead and wounded. The artillery continued to plaster the surrounding jungle in an attempt to kill as many as possible.

I lay in the foxhole thinking of Kendel. At times, I had grandiose thoughts of someone going with me and we could go out and find him. We listened for any sound that might have been him but heard nothing.

As the skies began to get light, a squad formed and headed towards the listening post. They found Kendel still leaning against the same tree he had been at when the attack began. His hands were raised in a grotesque attempt to ward off the danger. He was riddled from his neck to his abdomen with machine gun bullets. At least he had died instantly.

The squad also noticed something else. The listening post had made a deadly error. They had returned to the same spot as the observation post had been. During the time between the two, the enemy had taken vines from the trees, and tied them to use as hand railings. They quietly crept up to the end of the vines and knew they were just ten feet in front of the listening post. All they had to do was open fire. We were fortunate to get the other two guys back at all.

I returned to the top of the hill to see if Larry had lived through the night. It had been a rough night for him. There had been moments when he would come to and try to get up and run but would slip once more into the coma. We had called for a chopper to come at first light, and in the distance, I could hear the muffled sounds of the medevac as it slipped through the moisture laden air. As long as the cloud ceiling remained where it was, and we could keep the snipers away, the chopper could land, giving Larry and the other guy the chance to live.

I looked towards the sound and saw three choppers coming in fast. As they neared, two split off and I could see they were gunships sent along for protection. The medevac approached the landing area in a steep, fast decent. I saw one of the door gunners, standing on the strut, firing at the ridge they had just crossed. Then they were on the ground.

As soon as they touched down we were running forward with our two WIA's and the one KIA. All the while the door gunners hovered over their M-60 machine guns, and the gunships circled overhead, as we loaded the wounded on the Huey. The rotors began to turn faster and faster as the pilot prepared for lift-off. I touched Larry's hand, and looked into his eyes, but there was nothing. He had slipped into a coma once more. The chopper lifted to a hover, and once the pilot felt sure the load was not too heavy he increased the pitch of the main rotor and was on his way to the 18th Surgical Hospital in Pleiku.

I stood there and watched the chopper as it became a dot on the horizon and was soon out of sight. I turned and looked out over the field of fire in front of our perimeter and saw the new bodies that had not been there when darkness had come the night before.

As I stood there, my friends from weapons platoon came up to me and one of them asked if I was all right. Then another one said they couldn't believe what I had done the night before. They had watched as I had gone outside the perimeter to get the wounded guy, and had known both of us were going to be killed before we could get back in. They had then realized the enemy was not firing on that section of the perimeter at all. My friend Bob Stockton said that if I was ever in Chicago he was going to buy me the biggest steak that city had. As we stood there looking at each other, the Lieutenant walked up and told me he was glad nothing had happened. He too had watched and expected the enemy to open up and kill us both. He then said he was going to put me in for a Bronze Star.

I went looking for Harold and found him sitting by his foxhole looking like I felt. I never said anything, just sat down beside him. What had seemed like hours last night had lasted only a little over fifty minutes. It is not easy to gauge time accurately when death is walking beside you. We sat there gazing out in front of us, watching the guys inspecting the bodies, and finally decided to look, too.

I walked out into the killing zone in front of our positions, and there, among a dozen or more bodies covered with giant black flies, and maggots already eating away at the flesh, I realized I was witnessing a real war. Not something far away or something I was not involved with. What I was involved with was the senseless destruction of human lives.

I knelt beside one of the bodies, which only hours before had been a young North Vietnamese. I looked at his face and saw the horror of violence, not the peace he should have had. I opened his pack and saw the plastic shaving gear. The kind you would buy your son at the Five and Dime store. A letter half finished, and pictures of a young mother holding a child. He had a spoon and fork he had made from a piece of aluminum.

I was looking through a dead man's earthly possessions. A dead man whose family would never know what happened to him. They would only know that he never returned from the war. I moved from one body to the next. The type of wounds varied. Some were missing limbs and others where hardly recognizable as once being healthy young men.

We were tired, and both felt we would rather be somewhere else. Nothing much had changed since yesterday. We were still surrounded on a hill few people knew about, and fewer still would ever hear about. A hill that meant nothing to anyone. A hill we would hopefully walk away from soon, never to return.

Chapter Eleven

"Life is ten percent what happens to you and ninety percent how you respond to it."

—Charles Swindoll

Throughout the day, we tried to cover the bloating bodies of the dead NVA as best we could. We knew we weren't going to get rid of the smell, but we were willing to try anything. All the time we had to be careful because of snipers. Choppers came on a regular basis all day long and by evening we had all the supplies we needed. Empty ammo boxes were everywhere.

The medic from Bravo Company that had been in Base Camp showed up on one of the choppers, so we were finally up to full strength. I moved in with Harold, and we tried to enlarge his foxhole to fit both of us. It was mostly rock, so we weren't very successful. He is a couple of inches shorter than I am so when we both got in the foxhole we had to lay on our sides and I would have to bend my knees a little. We positioned logs and sandbags over the top of us, leaving a small opening to get out. This gave us added protection from grenades and shrapnel. The enemy must have run out of mortars because we had not had any more hit our perimeter since the first night.

During the late afternoon, Harold and I had a chance to relax. As we sat on the edge of our position talking, I noticed for the first time just how close the bodies were. Harold said the NVA must have been very close to them as they tried to get to high ground the first day. No sooner had they arrived on the hill and positioned the observation posts, they were hit.

The OP directly in front of where he had begun to dig his foxhole was immediately overrun. He looked up and saw two of the three guys from the OP trying to make it to the perimeter. One of the two was trying to drag the other. It looked like the guy's leg had more than one knee joint. The third guy was also wounded and was being held captive by the NVA just inside the tree line. The enemy had positioned themselves around him and every so often would poke him with a stick or their AK-47's. When he began calling for help, Harold and two others started out of the perimeter to get him. Suddenly Harold found himself lying flat on his back. His first thought was he had been hit. He looked up and he saw 46 year old platoon sergeant Charles Turner he realized he had been tackled. Sgt. Turner told them to stay where they were and taking off his shirt he began to fill it with grenades and M-16 ammo. Lt. Hunter saw what was happening and took his shirt off and began doing the same thing.

The NVA had positioned a machine gun just inside the trees about fifty feet away from where Harold was laying. They had expected someone to come directly out of the perimeter and try to rescue him. Instead Sgt. Turner and Lt. Hunter did an end run and came in behind them, catching the NVA completely by surprise.

With Sgt. Turner leading the way, they attacked with grenades, grabbed the wounded guy, and made it into the perimeter without being injured. On the way in they had knocked out the machine gun position which was just in front of Harold.

Harold had been trying to work on the guy with the wounded leg and now that the machine gun was out of commission he was able to examine the wounds a little closer. There were numerous bullet wounds plus all three major bones in his left leg appeared to be shattered. He was losing a lot of blood and it was important to get the bleeding stopped.

After giving a shot of morphine, Harold positioned the guy half over himself and was using his own legs as a brace. He was doing this while lying on his back. Whenever he would try to rise, another enemy machine gun would fire a burst over them.

Directly in front of Harold, and maybe seventy-five feet away, lay the machine gun that Sgt. Turner had knocked out. In a desperate attempt to retrieve the gun, the NVA sent one after another out to get it. In the end, seventeen had tried, and all seventeen had died.

They were the pile of bodies that now lay just in front of our position. It appeared they had been stacked there but they had just fallen over one another as they died. We looked at them and wondered how many more would die before we were able to get off the hill.

Small patrols had gone out during the day to see what was in the jungle around us. They could only get a couple of hundred meters out before they would begin to receive sniper fire, so we knew we were still surrounded. On their way back into the perimeter, one of the patrols captured an NVA sergeant. During the interrogation, he claimed he had gotten separated from his patrol and said that the NVA were going to try to overrun us. More soldiers were on their way from a camp just across the Cambodian border. He said a lot of his men had been killed and they wanted to pay us back.

These patrols had found bodies tied to bamboo poles which allowed the enemy to carry their dead away. A lot of broken weapons littered the jungle surrounding our perimeter, and blood trails indicated the enemy had taken a terrible loss.

We had been very fortunate the night before and we knew it. We were filled with apprehension as we thought about what the enemy prisoner had told us and of the darkness which was just a couple of hours away. When sunset came, we were as ready as we could possibly be.

Puff the Magic Dragon or Spooky was the names we called the AC-47 gunships. They had been modified by mounting three 7.62 mm General Electric miniguns to fire through two rear window openings and the side cargo door, all on the left side of the aircraft, to provide close air support for ground troops. The aircraft also carried flares it could drop to illuminate the battleground.

At 2000, Puff showed up to keep us company. He could stay around for a couple of hours at a time, but he assured us that when he had to leave another Puff would be there to take his place.

At 2030, the listening post directly in front of us began reporting sounds of movement. Their whispered voices indicated that they were extremely frightened and rightfully so. Without warning, all three of them came running into the perimeter. They were very fortunate that our own guns had not killed them as they ran in.

The procedure for the listening post was simple. They were to stay out until given permission to return or were hit by the enemy.

When they did leave their post, they were to blow the claymores and bring the radio and their weapons with them. We now had a slight problem in that two of the three had left their weapons at the listening post and the radio and claymores were still there as well. All three of them were now sitting beside our foxhole and were as frightened as anyone could get.

Sgt. Turner who had fought in WWII, Korea, and now here in Vietnam came over and sat down beside them and asked what had happened. After listening to their explanation, he reminded them that they would have to go back to their position and get the things they had left behind. The radio especially could not fall into the enemy's hands. If that happened, they could then listen to all our radio communication. He stood up grabbed his M-16 and said he was going with them. It was evident that no one wanted to go but they did not have much choice. What Sgt. Turner had said made sense.

As they snuck out of the perimeter once again, I wondered if they would live to see the sunrise in the morning. Within a few minutes the radio came to life and we heard Sgt. Turner's voice. They had found the listening post and were going to stay there for a while longer.

A few minutes later, we heard a voice ask what time it was. Someone said, "Why, who wants to know" and the voice responded, "Never mind GI, what time is it?" The problem with this exchange was the voice was coming from somewhere in front of our perimeter. Just then an NVA soldier jumped up and opened fire. He had crawled to within thirty feet of a machine gun position without being detected. He was able to get a couple of rounds off before the machine gun killed him. This was unnerving to say the least. The clouds had disappeared, and we had a full moon shining and we still had not seen any movement. At this point, we pulled our listening posts in and prepared to fight.

Puff reported seeing lights coming towards us from three directions. He asked if we could illuminate the perimeter in some way, and he would begin firing his miniguns. With all the empty ammo boxes, it wasn't hard to get some fires going. We had just gotten them started when he told us to get down, the enemy was almost on top of us. He no sooner got the words out of his mouth when all hell broke loose.

Harold and I lay in our hole watching Puff circle our perimeter five hundred feet above us. The miniguns buzzed like angry bees as lines of red came from them, the shells landing just inside the tree line. Every fifth bullet was a tracer and a gun that can fire fifty-five hundred shells a minute can put on quite a show. The parachute flares cast an eerie glow over us.

All our claymores and trip flares were used up in the first minutes of the battle. Our perimeter put out a steady flow of fire for the next fifteen minutes. Every once in a while, Puff would get out of the way and the fire support base would saturate the jungle around us with 105mm and 155mm artillery shells.

We laid in our foxhole waiting for a call for a medic, but none came. In the beginning the tree line had sparkled with the muzzle bursts of AK-47's, grenades, and B-40 rockets. As we lay in our hole, we could hear bullets hitting the logs and sandbags that lay over us, and we could clearly hear shrapnel from our artillery as it sliced across the hill.

When the firing ended, Harold and I made a quick check of the perimeter and found that no one had been injured. Even with Puff there to help us, it had been an intense battle. It was hard to believe that no one had been hit.

When we returned from checking each position, I grabbed my poncho liner and stood up to shake it out. Just then an artillery round landed close to the perimeter. When I heard it coming in loud and fast, I dropped to the ground. When I fell, I landed on a small stump, hitting me in the middle of my chest and knocking the wind out of me. Harold was sure I had been hit with shrapnel, so he jumped out of the hole and asked me where I had been hit. Because I couldn't breathe yet I couldn't tell him I had not been hit. He rolled me over and began checking for a wound. At this point I wanted to laugh and cry at the same time. When I was able to tell him what happened, we laid there on the ground and laughed till tears ran down our face.

Next to us was a three-man position. It was made so they could kneel with their arms resting on the firing ports. At 0300 the guy in the middle left to get more grenades. While he was gone an artillery round exploded somewhere out in front of us and we could hear something swishing through the air towards us. It entered the foxhole and imbedded in the wall right where he would have been kneeling. It was a six-inch, white-hot, jagged piece of shrapnel from the 105mm round. If he had been there, I'm sure it would have killed him.

As morning came we began to stretch our stiff bodies. A lot of us were extremely tired by this time. Three days without much sleep was beginning to take its toll. The Captain put us on half alert and the rest tried to get some sleep before it got too hot.

We were sitting by our foxhole talking and watching the choppers come and go when I noticed what looked like a man hiding in a tall tree across the valley half a mile away. Whenever a chopper would come, he would go to the back side of the tree then reappear when it took off. I asked the artillery Lieutenant if I could use his binoculars, and sure enough there was a man in the tree. I handed the binoculars back to the Lieutenant and suggested he look. After seeing the enemy, he got on the radio and called the fire support base.

The artillery was just getting ready to fire when we saw a Phantom jet fly by with one bomb hanging under a wing. The Lieutenant asked the artillery to hold for a minute while he turned to a different frequency and was soon talking with the pilot. Yes, it was an extra bomb and yes, he would be more than willing to circle back and drop it on a tree.

As he circled off in the distance, the artillery put a spotter round where we wanted the bomb dropped and then the Phantom began to dive. What a sight to see this jet come falling from the sky then level off and put his five hundred pounder at the base of the tree. When the dust settled, the tree was gone.

A patrol headed across the valley to check things out. When they returned, we found that the NVA had been digging new mortar pits. The bomb had destroyed the position and the ammo as well. No bodies were found but that was not unusual.

Around noon a chopper landed and out stepped a news team from NBC. As that chopper took off, another landed and out stepped five high ranking officers. As the news team began taking their pictures and getting their interviews, the officers began mingling with us. They seemed to be taking our emotional and psychological temperature. One of these officers sat down beside Harold and asked if there was anything he could do for him. Without hesitating Harold told him he could get us off the hill.

Since the first day on the hill, Bravo Company's CO Captain Alfred Jones had begun calling Sgt. Turner, "Poppa T" and with great respect for the man, the name had stuck. The Captain told the reporter how "Poppa T" and Lt. Hunter had saved the guy the first night, and how he had gone out to the listening post the second night. The cameraman took pictures of the battlefield and of us guys who had put up such a successful fight, then they climbed on their chopper and left.

We had gone all day without a shot being fired at us and as darkness came we hoped we had seen the last of the enemy. Throughout the night, we could only hear the normal jungle noises and the listening posts reported no movement at all. Puff returned around 2100 but things were so quiet he finally left and went back to Pleiku.

When the sun came up, we were told to prepare to leave. We were going to try walking off the hill. The perimeter was not large enough for an LZ, so we had no choice but to walk away. This brought mixed emotions because the perimeter had been rather safe after the first night. It afforded us a feeling of security which we were not willing to give up easily. As soon as we left its safety, we would become vulnerable once again.

We sent a squad size point out ahead of us and cautiously started walking away from the hill. I began to realize that Harold and I had just experienced a major transition in our lives. With God's help, we had shared an experience together that would last a lifetime. We had defeated death, God had won, and if we chose, He could use us once more.

It wasn't long until we were deep in the jungle. The smell of rotting flesh still hung in the air but was now less offensive. The tension was high as we prayed that the NVA had had enough and would leave us alone. The heat and humidity hung in the air almost visibly. It was like walking through a steam bath wrapped in a hot towel. Our fatigues were soaked with sweat, turning them black. There was no relief from the heat, no breeze blew to cool us.

We worked our way deeper into the twilight of the jungle. Occasionally patches of sunlight could be seen where there were tiny breaks in the thick canopy a hundred feet above. The deeper we went, the dimmer it became until we were in a world of perpetual twilight. The jungle thinned abruptly, the broad-leafed plants and lacy ferns giving way to thick grass several feet tall.

We walked alone in single file with only our thoughts of the past five days. Memories of dead NVA left rotting on the jungle floor and of our friends who were wounded, never to be the same again, and of our friends who died there.

After an hour or so, the pace began to take its toll on us and we began to slow down. We halted and fanned out in a rough circle. We hadn't been followed and it appeared the enemy was going to leave us alone. The tepid water in my canteen tasted good, even with the halazone tablets in it. The insects attracted by the sweat darted at our eyes and buzzed around our ears. After a few minutes, the monkeys began their usual chatter and the birds squawked. These were good signs. We could rest easier because we were alone.

Half an hour later, we saddled up and moved out once more. The point and flank security cautiously set out and we followed. In a few minutes, we topped over the crest of a small hill and down into the valley where a stream flowed. The ground turned marshy and wet, and the water felt cool and refreshing as we crossed the small stream. For once I wished it would have been deeper and we had time for a bath. Harold had been telling me a bath would do us both a world of good.

We walked most of the day and then entered an old deserted

village just before sundown. At the far end of this village was a piece of land sticking out into a deep rugged valley like a finger pointing west. On this finger of land, we found Charlie Company waiting for us.

A perimeter had already been formed and we were to spend the night with them. On the three sides of this finger was steep shale, and the land was covered with giant teak trees. At the end of the finger was a log platform which they had built to be used as a landing pad for choppers. The Hueys would enter the valley and touch down on this pad one at a time, their rotor blades coming very close to the trees. When they were ready to take-off, they would have to lift off slowly, turn around, and leave the same way they had entered.

Charlie Company had also been in contact the past few days and this landing pad was built out of a desperate need for resupply and evacuating the dead and wounded. The two engineers that traveled with them had performed an almost impossible task. With the help of Charlie Company, they had built this platform by hand. The logs were huge and it was built out over the side of a steep hill. One person had been killed when a tree fell on him.

To get to this finger of land, we had to cross over an open area about thirty feet wide and a hundred feet long. They had positioned their two machine guns at this point to keep the enemy from entering. The village was deserted but some of the animals were still there. It appeared to have been a Montagnard village and the people had fled.

Throughout the night, the enemy would sneak through the village to get as close as possible to Charlie Company. As they came, the chickens and pigs would become frightened and start making noise. Because of this they had known the NVA were getting close. They had been able to repulse every attack and, like the enemy that had been trying to get at us, they too had fled back into Cambodia.

As we entered the perimeter, I saw Ruben sitting in one of the positions. It was good seeing him. We all had been so close during our stay at Ft. Lewis and when we arrived in Vietnam we had become scattered in separate units. Out of our group I was only around Corky while we were in the field. Ruben looked as tired as I felt. The walk had been filled with stress and the humidity had been draining. At least now the guard duty would be shared between our two companies, and most everyone could get some much-needed rest.

I looked at the luminous dial of my watch to see what time it was. It was still too early to get up, so I lay in silence listening to the nocturnal noises around me. They were safe and comforting sounds to hear. Soon dawn would be breaking, and the blackness would be turning to shades of grey. I looked to my left and saw the faint outline of the next position and could make out a dark lump of the person's head. I let my eyes wander over the jungle, listening to the light rustling of leaves as a breeze blew through. Soon I could pick out individual trees and bushes as the jungle became light shades of grey instead of black. Patches of mist drifted up from the valley below and the sky became a golden ball of orange. Another humid day was about to be born.

My thoughts drifted back to the day I came ashore on the beach at Qui Nhon. What a contrast that was to where I found myself now. When our C-130 had circled out over the ocean and headed inland towards Pleiku, I knew I was going to miss those lazy breakers as they caressed the white sandy beaches of the coast. An hour or so later, we landed in Pleiku and the cultural shock began. What little we had seen in Qui Nhon was now magnified by the sights of rural Pleiku.

I looked up towards the triple canopy a hundred feet above where I lay and wished I could be back on those white sandy beaches listening to the waves. Anywhere would beat what we had experienced the past five days.

Hemingway once said, "There is no hunting like the hunting of men and those who have hunted armed men long enough and liked it never care for anything else thereafter...." All of us inside that perimeter had experienced something new and tragic over the past five days. I doubt if there were very many who had enjoyed the violence that had and still did surround us.

Chapter Twelve

"I said to the man who stood at the gate of the year: 'Give me a light that I may tread safely into the unknown.' And he replied: 'Go out into the darkness and put your hand into the hand of God. That shall be to you better than a light and safer than a known way.'"

—M. Louise Haskins

Bravo Company stayed with us for the next two days, as we joined forces and patrolled along the Cambodian border. Our fear of being attacked lingered with us as we moved through the jungle. Our senses were stronger now and we no longer took chances like we had before the battle began six days ago. At night, our foxholes were not the impression in the ground they had once been, but a position worthy of a veteran. They were dug by boys who had become men over night. Men who now understood and valued the life they had, regardless of the problems it might hold for them.

Around noon we came to the Ia Drang river. The water was swift, and the river was at least a hundred and fifty feet across. One of our engineers said he was a pretty good swimmer and would be willing to swim across with a rope. Lacking a better plan, we tied a rope around his waist and he dove in. Half way across, the current pulled him under and when he didn't surface we quickly began to pull him back to shore.

I had been watching when this happened and when he had gone under I headed for the water's edge in case he needed medical help once we got him ashore. I stood by the water brushing up against a bush that was about five feet high. I heard someone say, "Don't move, Doc, there is a snake right behind you ready to strike." Every snake story I had ever heard flashed through my mind as I froze in place. The memory of the 22-cal. pistol exploding in my hand went through my mind and I remembered the peace it had brought me when I had realized it was God's way of letting me know that everything would be all right.

I looked over my shoulder and saw the unmistakable red stripe on the tip of a very green tail. Just then a machete slashed through the air and the crisis was over. Someone reached into the bush and brought out a four-foot Bamboo Pit Viper. I never had time to collapse although I felt like I wanted to because the swimmer was just coming out of the water.

When the current had taken him under, the rope became tangled around his legs. When he tried to free himself the guys on shore began pulling him in and the tension on the rope made it impossible to get free. He had no choice but to hold his breath and enjoy the ride back to shore. Other than needing room to breathe, he was alright.

There had to be a better way to cross. All we had to do was find it. Captain Summers sent a patrol up river to see what was there. Fifteen minutes later the radio came to life and the patrol said they were just around the bend at a natural crossing which was only about a foot deep.

We had wasted an hour, but the stop and the stupidity of

what we had experienced was maybe just what we needed. We had been so tense from the past few days and this had helped to take our mind off the enemy. We continued on, more relaxed yet cautious. We had seen no signs of the enemy, but we knew he was still sharing the same jungle with us.

We stopped for the night around 1600. Earlier that afternoon we found a well-used trail. The plan was to send a six-man patrol to set up a night ambush and see what might happen and according to Lt. Underwood I was to be the sixth man.

That evening while the world was ghostly still around us we slipped silently out of the perimeter and headed up a ravine. We had about a third of a mile to go before we would turn into the jungle and go another quarter mile to the trail. We were about ready to make our turn into the jungle when we noticed two NVA following us just inside the tree line. We opened up on them and headed up the bank.

As we were climbing the bank I heard Sgt. Moran swearing. I turned and asked him what was wrong. He said he had dropped the pin to a grenade and couldn't find it. If he didn't drop it, we were safe. I wished him good luck and headed the other direction.

We lay tense and breathing quietly among the thick undergrowth at the edge of the jungle. The moon had not yet risen high enough to flood its light around us. We moved another 100 meters and set up a defensive perimeter. Lt. Underwood and I lay in the center and the other four lay down with their feet touching us. No one was to sleep, and we were not to fire unless the enemy was on top of us. Even throwing grenades was dangerous because we were not sure where the trees might be.

This was not my idea of a fun way to spend the night. We

had been there only a short time when we began to hear voices and movement. The NVA was trying to flush us out. All it would take was a muzzle burst or a cough. They might even be able to stumble upon us by accident. It sounded like they were a couple hundred feet to our right. Before we realized it, the jungle blackness was turning to gray.

Then in the pregnant silence of the brink of day we saw we were surrounded by a crisscross of bamboo. It was as if all life held its breath. The very utterances of silence shouted in our ears as we crept out of the jungle and headed towards the safety of Alpha Company's perimeter. As we neared the perimeter, we called to let them know we were close. As soon as they told us to come we took off at a full run. I doubt if I had ever felt as good as I did when we were inside the perimeter once more. I told Lt. Underwood that if he ever told me to go on another night ambush with anything less than a full company I would probably tell him where he could go. He said he doubted if he would ever be as bold again as to volunteer our services.

For the next week, we walked the hills and valleys of the Ia Drang Valley. The first couple of days we found bodies and weapons left by the NVA we had fought. Their trail clearly led over the river and straight into Cambodia. Our intelligence had a good idea of the route they would take and pressure from patrols, artillery, and the Air Force continued to take its toll on them.

We found another shallow part of the river and crossed back to the other side. We knew we had been close to Cambodia the night before. We had been walking all morning and seen nothing but an occasional glimpse of the Ia Drang river. Around 1000 we found ourselves under a triple canopied ceiling. The undergrowth was thick, and the going was slowed by the tangle of bushes.

Pushing away branches and vines, we found ourselves standing in what looked like a living green tunnel. The trees had been pulled together at the tops and tied with vines. The ground was packed hard and resembled a one lane paved road. We were standing on part of the Ho Chi Minh Trail. From the air, even the trained eye could not detect what was on the ground. An army of men and equipment could travel without fear of detection as they walked east from Cambodia into Vietnam. Pleiku City was only forty miles from the Cambodian border.

The Captain relayed the grid coordinate back to battalion headquarters so the LLRP's could keep an eye on it and we continued north. The artillery needed to move forward, and with the discovery of the trail it was decided to leapfrog past this point and build a new fire support base. Word came about 1230 that we needed to be at a large LZ not far away in about an hour. We were going to be flown north over a small range of mountains. This was the best news we had heard in a long time.

Just as quick as good news can come it can go. As we were hot footing it to the LZ we received new instructions but this time it was even better news. It had been decided that we would be taken a little farther north to a fire support base that was already operating and take over security for a week. We soon arrived at the LZ and set up a perimeter to secure it. In the distance we could hear the popping of rotor blades as eighteen Hueys with their escort of gunships made their way through the moisture laden air to get us. I looked to the west and saw the low-hanging clouds moving our direction promising that rain would be with us soon.

Towards the north, the sun was bright and looked inviting. Maybe we would be able to get clean clothes and get dried out.

Two stress filled weeks of walking in the jungle had been enough; we were ready to go in for a rest.

I picked up my pack and eased the straps over my shoulders, buckled them and adjusted it to ride high on my back. Even though I had less weight now than when I had walked off the hill two weeks ago, it was still heavy. I was tired and ready for the relative safety of the fire support base.

Security was in place and the rest of us were in our groups of six staggered across the LZ. I was standing next to the radio operator and was listening to his conversation with the choppers. He nodded to the Lieutenant who took a green smoke grenade and tossed it in the middle of the LZ. Over the radio I heard the lead pilot say, "Amigo Six, this is Peacekeeper Leader, I have Green smoke in sight." The RTO answered and told him it was our smoke and the LZ was cold. It was safe at the moment to come on in.

Only those who have seen it could understand what a beautiful sight a large flight of Hueys can be, especially if you are their destination. They came into the LZ in groups of six. The pilots flaring slowing their forward momentum and then settling into the foot-high grass. We quickly climbed aboard and took our places on the web seats or on the floor.

The pilots stayed on the ground only long enough for us to get in. The engine noise increased, and the choppers lifted to a hover, hanging in the air five or six feet above the ground. Suddenly the nose dipped, and we were racing across the LZ towards the tree line. Up we popped over the trees reaching for altitude. The NVA loved to catch the Hueys at such a low level and so vulnerable. Soon we were at fifteen hundred feet and on our way to the fire support base a twenty-minute ride away.

The week guarding the artillery was very relaxing for us. I

found I could sleep even with the 105's firing. We were located on the top of a small mountain overlooking the Cambodian border. The sunsets were beautiful, and the nights were so clear you could write letters by moonlight. A week later we prepared to move out once more and Bravo Company would take their turn as security for the fire support base.

Three days into our patrol we received word to secure an LZ which was a good four-hour hike away. We needed to be at the LZ, and have it secured in three hours so off we went. Luckily the terrain was fairly flat, but we would still have to keep up a swift pace. I did not like moving this way because our awareness of what was around us slipped when we moved fast.

We had been going for a couple of hours and were drained from the humidity and heat when we began to smell death. Somewhere close by there were men laying in the jungle dead, the heat baking the bodies. The tropical heat and humidity added to the process of decay. The trees around us showed signs of violence. Trees uprooted and cut in half by giant artillery shells. And since the death of men were violent, their bodies ripped apart by shrapnel, exposing the soft interiors the decay spread rapidly. Ahead along the trail I noticed a body lying beside a tree that had been hit by an artillery shell. As I walked past, I could see he did not have a head.

We were about fifteen minutes away from the LZ when we noticed old fortified positions all around us. They were so well hidden that we were inside their perimeter before we noticed them. We quickly checked a couple out and when nothing was found we continued on.

Shortly after we secured the LZ the Hueys began to arrive. Over the next hour, we watched as several hundred Montagnard

Strikers arrived along with their Special Forces advisers. As they would land, these little warriors would form up in platoon or company size units and melt into the jungle. Most of these Strikers looked to be younger than we were.

The last company to get there headed down the same trail we had used just an hour or so earlier. They had been gone only a short time when the jungle resounded with gunfire. We had broken the perimeter and were ready to head on out when this happened. We quickly reorganized a tight perimeter and waited. Captain Carlson, the SF adviser, called on the radio and asked if we could send a platoon to secure the trail between our perimeter and the fight. He had wounded and wanted to send them back into our LZ. A platoon saddled up and quickly headed out.

I checked my aid bag and stood waiting for the first casualty to show up. Soon I saw one of our guys heading in with a Montagnard who couldn't have been over fifteen years old. In his left hand, he carried an old Garand M-1 rifle and his right hand was being supported by the rifle sling. He walked up to me and I noticed beads of perspiration all over his face. Smiling at me he held up his right arm. About six inches of muscle between his wrist and elbow was exposed. The wound caused by fragments from an enemy grenade. As I began to put a sterile dressing on I wondered if I should give him a shot of morphine or keep it for someone who may need it later. Here was a kid standing in front of me with his muscle exposed, obviously in pain, but politely smiling at me. I had worked on my fellow Americans with less extensive wounds who needed morphine just to keep them quiet. I decided to wait on the morphine. When I finished dressing his wound I showed him a place in the shade where he could sit down.

We soon had eight little Montagnard sitting around with various wounds. The enemy broke contact and melted back into the jungle. Word came in over the radio that the Strikers and our platoon were on their way back in. As they entered the perimeter I noticed the two Special Forces advisers carrying the native Lieutenant on a bamboo stretcher. He had received a stomach wound and seemed to be in a lot of pain.

He lay on the ground moaning loudly like he was about to breath his last. I bent over him and lifted the bandage away from the wound and saw two little holes made from an AK-47. Capt. Carlson told me his moaning was mostly for show. He had gotten himself wounded and therefore been disgraced in front of his men. The little warriors were standing around him telling the Captain all the Lieutenant needed was a Mjao.

We had medevac choppers on the way and soon these brave fighters would be in a hospital lying between clean white sheets. Just the thought of clean white sheets was reason enough to me to justify a small wound.

Throughout my high school years, I had dealt with the problem of weak ankles and always seemed to be on crutches because of a basketball injury. I had begun to wish for a broken ankle or even malaria, just so I could get back to civilization for a while.

After the medevacs had left, we teamed up with the Strikers and headed towards the border. We stopped early that afternoon and set up a good perimeter. We were at the crest of a small hill that had a good size stream maybe half a mile away. The Montagnard's formed the first line of defense and we made up the second.

It was a new experience for us being as we had only seen Montagnards from a distance. We watched as they set up camp and began to fix their meal. One of them had caught a lizard that afternoon and had been carrying it tied around his neck. We now understood what it was going to be used for. We would complain about having to eat our C-rations but after watching them sit around and devour their rice and fish heads with Nuoc Nam Sauce with chunks of lizard on the side we came to a better understanding about good food.

As darkness came we began to turn in for a decent night's sleep. With the Strikers having their perimeter around ours we knew we could sleep sounder than usual. At 0200 I felt something nudge my shoulder. I opened my eyes and looked into the face of one of the Strikers and their interpreter. It seemed that one of his friends was suffering from a high temperature. I took my aid bag and followed them to the other side of the perimeter.

The moon was so bright I didn't need a flashlight to see that the thermometer was showing a temperature of over 104 degrees. I woke Capt. Carlson up and asked him for a squad of his men so I could take the sick Montagnard down the hill to the stream and try to get his fever down.

Within minutes we were ready to leave. We snuck out of the perimeter and were soon by the water's edge. As soon as we arrived, the squad of Strikers disappeared into the darkness and set up a perimeter around us. I told the interpreter to have the guy strip and get in the water up to his chin. After twenty minutes of listening to his teeth chatter the fever was gone and his temperature was back to normal.

When he had gotten dressed he turned toward me and put his hand on my shoulder. He said something in his native dialect and took his bracelet off and put it on my wrist. I looked at the interpreter and he told me the Striker had said that he would forever be my friend and the bracelet was to be a token of this friendship. We stood there in the darkness with only the soft sound of the stream and looked at each other. I was at a loss for words. I was glad it was dark so this little Montagnard warrior could not see the tears trickling down my face.

When morning came we called for a medevac and my new friend headed for the hospital in Pleiku to recover from a bout of malaria. We had decided to stay in the perimeter for the day as patrols searched the jungle around us.

Throughout the morning, I noticed some of the Montagnard's would walk past my position and casually look my way. When our eyes would meet, they would smile and walk away. Most of the strikers had the same thing in common, they all needed a haircut. When I had left Base Camp, I had taken the aid stations hand clippers with me. I had used them frequently over the past few weeks and had become pretty good at cutting hair.

I looked around until I found one of my friends who needed a haircut then positioned us where we could be seen by the Strikers who were still in camp. Before the day was over I had given over twenty haircuts and at supper time I was invited to share their meal.

I just couldn't force myself to look a fish head in the eye then chew it ten times like I had been taught and swallow what was left. Luckily for me these guys also had Special Forces survival rations so using the interpreter I told them I had always wanted to eat one of these survival rations. After some discussion, they decided that if I ate one of the survival packs then they would like to try some of my C-rations. That was an equal exchange I was all too happy to make.

The next morning, we pulled camp and headed out again. Around 1130 we found a small NVA campsite. There were several dilapidated thatched structures that kept out the monsoon rains but not the things that crept and crawled throughout the jungle. It was interesting watching the Strikers as they went through the camp with a fine-tooth comb. It was determined that the camp had been used the night before and would more than likely be used again soon.

As we stood around deciding what the plan of action was going to be, up the trail walked two NVA regulars with their AK-47's slung over their shoulder, obviously unaware that we were there. They recognized us about the same time we noticed they were not Montagnard. Fortunately for them they were a little faster than we were, but in their haste to leave one of them literally ran out of his Ho Chi Minh sandals leaving them on the trail in front of us. With their supply problems as it was I wondered how long he had to wait to get something else to wear for shoes.

The Montagnard's were hot to follow these two NVA, and our patrol was going to take us in the opposite direction, so we said good-by and headed out on our own. The plan was to join up again in a week or so. Over the next two weeks the Strikers we had seen fly into the LZ would take over forty percent casualties. That night as we set up our perimeter we received a radio message that three other units around us were getting hit. Even the fire support base was being hit by mortars and a ground assault. We quickly dug our holes a little deeper and as the sun set our LP's snuck out and set up their post. They soon began reporting movement on every side. One LP reported hearing something land close to their position. When it got light they found a grenade that had not exploded only a couple of feet in front of them.

We were soon to hear that we had been the only unit in our AO who had not been attacked. The fire support base had taken six killed and twenty-two wounded. We had been the lucky ones that night. The NVA had known where we were but apparently did not have enough men to attack everyone.

Montagnard Bracelet

Chapter Thirteen

"People sleep peaceably in their beds at night only because rough men stand ready to do violence on their behalf."
— George Orwell

November saw the third of our medic group wounded in action. When we arrived from Ft. Sam, Charles White from Montana, had been assigned to Alpha Company 1st Battalion 22nd Infantry. During their stay in Tuy Hoa guarding the rice harvest his unit began getting mortared one evening and had a wounded guy outside the perimeter a little bit in front of him. The artillery FO had requested artillery support and the first round had landed. Charles heard him say, "Drop 25" which was telling the artillery to bring the next round 25 meters closer to the perimeter. Charles jumped up to run get the wounded guy before the round landed and either an enemy mortar round or the artillery round landed in a tree and took an inch of bone from his right forearm.

In a letter to my folks written from the hospital dated June 10, 1967, he sent this report. "The doctors are a little troubled about my arm. It just doesn't want to heal so I am going to be a G.I. for many months yet. Oh well I am alive and have all my arms and legs, so I won't complain."

In an earlier letter to my folks, before he was wounded, he told a little about what it was like where he was.

"I had a great sleep last night. I was able to sleep in a hammock. It was quite saggy, but oh so comfortable. I am sitting on a rock pile under a bridge where it is cooler but up by the road it is so hot. I am able to take a bath every day and wash my clothes. The bugs are terrible and plentiful. You should see the lizards, they are more like alligators. There is also a lot of birds and we can often hear monkeys."

The month of November started out on a positive note for me. I was promoted to Spec. 4, and with it would come the pay increase bringing my monthly wage up to $240.00. The sad thing of getting PFC pay was the last 28 days of the month. Now that I was to be a Spec. 4 I was sure it would be better. It would now be the last 25 days. What a bummer.

We began to find more and more signs of the NVA's build-up in the Ia Drang Valley. Short skirmishes were becoming almost a daily event. The Montagnard strikers took a terrible beating during the early days of the month and the enemy showed they were serious by attacking the larger fire support bases as well.

The monsoon season was almost over, and the heat and humidity were becoming a real problem, especially for the new replacements. Even in the jungle under the double and sometimes triple canopy it was at times almost unbearable. Guys dropping from the heat were a common occurrence and we looked forward to the streams we would have to cross.

On November 10th, we arrived at a small fire support base that Bravo Company was guarding. It was late in the evening so not much time was left for us to dig in. The perimeter was well fortified already so we felt safe. Some of us took up positions on the perimeter and the rest settled in for the night. Another fire support base had fired rounds around our perimeter earlier that afternoon so if we needed support they would be ready. When we arrived, we found a Lt. Brannen waiting for us. He was a replacement for Lt. Underwood who was heading for Bangkok on R&R. When we settled down for the night we watched the Lieutenant as he spread out his poncho and then his poncho liner. This was nothing unusual except he then took off his boots, socks, and pants.

While he was neatly folding his pants and placing everything on his poncho, we were watching lights that were entering the clearing from three directions to our front. The NVA were completely unaware we were there. As soon as Lt. Brannen got everything neatly situated I said, "Hey Lieutenant, do you think it is a good idea to get undressed?" After giving me a couple of reasons, I pointed out the lights that were now about eight hundred feet away and asked if they might change his mind.

The first rounds of artillery were passing overhead when he began to lace his boots. Some of the NVA had already crossed a small stream and were walking up to our perimeter when the first shells hit. It was a dark night, so we could only see the lights. Most of these lights went out as soon as the artillery began but those farther away could be seen leaving the clearing for the safety of the surrounding jungle.

We let the artillery saturate the clearing and the jungle around us for a good ten minutes while we stayed down and never fired a shot. Noises outside the perimeter could be heard throughout the night. When morning came, we found that the NVA had cleaned up the battlefield. All the dead and wounded had been carried away.

We spent the next week patrolling the jungle around this clearing. On the evening of the 12th, a unit of the 1st Battalion 12th Infantry was guarding an artillery base about two miles from us. During the night, we could hear sounds of battle. Planes could be heard circling in the distance and at times diving in to drop bombs. The sound of Puff's miniguns was clearly audible but he was behind a hill, so we could not see him. We monitored their radio communications and by morning we already knew they had taken a beating. They were calling for medevacs throughout the night for their six KIA's, and twenty-two WIA's.

Around 0400 we received a message that the 1st of the 12th had captured an NVA prisoner. He told them about re-enforcements that would be crossing the Nam Sathay River that evening. The crossing spot was not far from us, so we saddled up and took off. As night fell we were in place and wide awake. The next ten hours or so could prove to be very interesting. As usual it didn't and by morning we realized we had spent a night with the bugs and anything else that may have crawled around us that we had been unaware of.

On our way back to the artillery base, we began to hear tremendous explosions coming from in front of us. During a re-supply of 105mm artillery shells a Chinook had caught its sling load on a tree and crashed just inside the perimeter on top of the ammo. Almost as soon as it crashed it burst into flames.

Art Collins ran into the burning wreck and brought the pilot out then returned and carried the co-pilot to the closest bunker he could find. By this time, the ammo under the flaming wreck was getting pretty hot. Recognizing the potential danger, most everyone had grabbed their weapons and set up a new perimeter across the clearing at the edge of the jungle.

It wasn't long before the shells began to explode. By the light from the burning wreck and the flashes from the explosions, Art realized he had brought the injured co-pilot into an ammo bunker full of 105mm shells. There was nothing he could do about it so while the Chinook was being destroyed just a few hundred feet away Art lay in the bunker and set the guys broken leg and prayed his hiding place would stay in one piece. When morning came, Art had his patient ready and waiting for evacuation. He was awarded an air medal for this act of bravery.

We spent the day helping put things back together and that night watched parachute flares exploding a few miles away and at times we could hear the muffled sounds of battle. When morning came, we saddled up and headed out to patrol the hills and valleys to the west. Recon platoon had set up an observation post on top of a hill just outside the firebase. As we passed through their perimeter I noticed most everyone was cleaning their weapons or reading magazines.

It was about 0800 and the coolness of the jungle was beginning to feel good. We were ten minutes past the observation post when our point rounded a blind corner and literally collided with three NVA. He opened up on them and saw one of them fall. When we got to where he had fallen the only thing found was a pool of blood and a brand-new Chinese machine gun.

Our morning began to deteriorate rapidly. We made the mistake of running down the trail after the three NVA. It wasn't long before the guy next to me was shot in the elbow by a sniper. I put a dressing on his wound and because there was not a clearing for a medevac to pick him up I took two guys with me and we took him back to the observation post.

Right after we rejoined the rest of the company we came to a ravine that looked like an excellent ambush sight. We spread out along the trail and let a platoon go through and secure the other side. As I stood there waiting, I saw someone in front of me take out his .45 to check the ammo clip. While he was checking it he accidentally pulled the trigger and shot the person next to him in the abdomen.

This was getting serious. A .45 is a powerful weapon and we needed to get a medevac in as soon as possible. The wound was high in the abdomen and we feared a possible sucking chest wound. While I dressed the wound, others built a bamboo stretcher and began carrying him in hopes of finding a clearing soon. Within a thousand feet or so we came out of the jungle into a valley that was about a third of a mile wide. It was filled with tall elephant grass and gradually dipped down a hundred feet or so from the level we were.

As soon as we came to the jungles edge we began receiving sniper fire from across the valley. In the distance, we could hear the pop of rotor blades as the medevac and its escort of two gunships approached. As the medevac circled a short distance behind us the two gunships made passes through the valley. As soon as one gunship made a pass while the door gunners hammered away with their M-60 machine guns the other would follow firing rockets into the tree line. After three passes, the medevac came in fast and we loaded the wounded and the three Hueys left.

Weapons platoon set up a perimeter to guard the small LZ and the rest of the company took off across the valley. As the last of the company was entering the elephant grass, I noticed what appeared to be fresh dirt. I walked over to it and found a hole approximately twelve feet square, and fifteen feet deep. This was a freshly dug mortar pit that was going to be used against us. The dirt beside it was still moist which indicated it was probably dug the night before. Something seemed strange though, and I could not put my finger on it.

Soon after they entered the elephant grass they came under machine gun fire from the NVA who were in fortified bunkers hidden in the grass. For the next five hours, they lay in the valley, pinned down by four machine guns. All the while fighting off small units trying to flank them.

The snipers had been silent since the Hueys had fired into the tree line and this was beginning to give us a false sense of security. I heard over the radio that two of our guys were down and caught in a cross fire. It wasn't long before we heard there were two more down, and that all four were confirmed KIA's. Soon after we made contact, a Cessna Bird Dog spotter plane began circling above us. Occasionally, we could see a line of tracers arching upwards as the pilot made a low pass over us.

We received word to secure the LZ because a Huey was going to be landing. We could see a chopper circling in the distance and watched as it began its approach. The door gunners were hammering away on their M-60's as the spent brass casings collected at their feet or spilled out the open door onto the battlefield.

The pilot held at a hover for just a moment. Just long enough for the two guys who were on board to jump out. After they were clear, the Huey began to slip forward. It held at about fifteen feet as it glided over the elephant grass gaining airspeed. As it neared the tree line we could see bullets striking its tail boom. The pilot pulled back on the cyclic and it climbed for altitude. Soon it was a speck its sound lost by distance and the noise of battle. Above us flew two A-1E Skyraiders waiting to drop their loads of bombs and napalm.

It wasn't long before we could hear airplanes approaching from the east. With a roar the two Skyraiders passed overhead at tree top level. As one banked to the left the other dove at the tree line. His pass took him parallel to the jungle's edge and we saw what appeared to be small objects falling. Soon the tree line erupted in bright flashes and the pop, pop, popping of explosions were heard. He was dropping cluster bombs which was meant to rid the trees of snipers. As he turned toward the sky the other Skyraider followed him through the valley dropping his load of cluster bombs at the valley's edge. All the while we could see tracers from AK-47's just missing his tail. They made more passes firing their machine guns and then they were gone.

For the next couple of hours, we received support from Huey gunships, the Skyraiders and our own artillery. Weapons platoon continued to guard the LZ while the rest of the company tried to advance. It became obvious they were not going to be able to take the bunkers, so they began to withdraw.

The first objective was to get the dead off the battlefield then try to break contact. It became clear they were not going to be able to get the bodies out. The cross fire from the fortified bunkers was too heavy. No matter how hard they tried, they were not going to be able to reach them.

So, the rest of the company began to disengage and started heading our way. As soon as the NVA realized what was going on they left their bunkers and swarmed into the valley. A murderous sound of gunfire erupted from somewhere in front of us as the first of the company entered our small perimeter.

Without warning, we heard artillery not passing overhead but landing on top of us. We only had time to fall to the ground as enemy shells landed in our small perimeter. I pressed myself into the ground in an attempt to get away from the death that flew amongst us. I opened my eyes and peered under my arm and saw bright flashes all around me.

As the last of the shells landed I lay there too stunned to move. I could hear moaning close by but felt I couldn't move. Once the shock was over, I began going from one person to another looking for the source of the moans. I soon found our company RTO lying face down, bleeding from a hole in his shoulder. Around me were fifteen others with minor wounds.

As I bent over the RTO I heard the airplanes returning. This time they had NVA in the open. One Skyraider made passes over the valley firing his machine guns to pin them down while the other got ready to come in from the other direction. I heard a tremendous roar overhead and looked up just in time to see a napalm canister fall from the wing of the Skyraider.

We realized the napalm canister was going to land very close to us. Time seemed to go by in slow motion. I watched as this large canister floated end over end, no more than two hundred feet above us. It passed our little perimeter and landed a couple hundred feet away but did not explode.

We grabbed our wounded and headed into the jungle. The air seemed to be electric. We went about five hundred feet into the jungle and set up a perimeter on a small incline.

Third platoon had deployed in a skirmish line facing the LZ we had just left. They were putting out a steady stream of fire, some firing on full-auto and some were shooting single-shot. They were putting up a good fight to keep the enemy away as we gathered our wounded in preparation to get back to the fire support base.

One of the wounded that came out of the valley was one of

our machine gunners. He was about 6'3" and weighed well over two hundred pounds. He had been hit in the thigh with an AK-47 round and it had shattered his femur. Even though he was in a lot of pain he was doing quite well. I had given him a shot of morphine and was trying to get the others ready for our dash to the fire support base.

I was kneeling over the machine gunner checking his wound when the Captain came to me and asked about his RTO. I told him I had been able to stop the bleeding and had given him a shot of morphine. The wound wasn't that bad, but he couldn't walk out, and we needed to make two stretchers. I needed some bamboo and two ponchos.

Two of the wounded we would have to carry plus we also had seventeen walking wounded with various wounds. I was soon given four lengths of bamboo. I laid them in the middle of the ponchos, about eighteen inches apart then folded the sides inward overlapping the bamboo. In just a few minutes we were ready to move out. I could hear the Captain calling the fire support base requesting that one of their platoons start up the trail to meet us.

I was reaching down to pick up one end of the stretcher that had the machine gunner on it when another round landed on top of us. Artillery and mortars had never bothered me in the past. We could always hear them coming and going over us. Even though it sounded like a freight train it had never been a threat. After the rounds that had hit our LZ it was not hard to tell when a shell was going to land close. Instead of it sounding like a freight train it resembled a fast, high pitched scream.

I fell over the machine gunner and the other three guys fell over me. The shell hit a tree maybe twenty feet away. Before we could reach the ground the other three guys had been hit by shrapnel. The one that had been standing next to me had been hit in the elbow and spun a good eight feet away. The other two had received small pieces of shrapnel in their arms.

As I took care of the three wounded around me I saw two other medics working on someone laying under the tree that had been hit. As soon as I could I went over to help and saw a guy laying there with his left arm missing just below the shoulder. I looked at the horror on his face and I began to get angry.

The NVA was pushing third platoon closer to us and the urgency to get going was mounting. We soon had everyone ready and we took off. We hadn't gone more than a thousand feet when we rounded a corner and found an NVA soldier sitting by the trail waiting for us. He had had enough and wanted us to capture him.

By this time, the enemy had broken contact and 3rd platoon was bringing up the rear. We soon contacted the platoon from Charlie Company. They let us pass and fell in behind us. Within the hour, we were safely inside the perimeter and medevacs were landing one after the other.

Right after the last medevac lifted off, a command chopper landed and off stepped a Lt. Colonel. He wore spit polished jump boots and fatigues that were starched so much that if he had been wounded he probably wouldn't have fallen over. He walked up to the prisoner and without saying a word decked him with a right cross. He then picked him up and threw him into the cargo bay of the chopper and climbed in behind him. Overhead flew a flight of Air Force jets and moments later we could feel the earth shake as they dropped their bombs on the fortified bunkers.

Sunset was not far off and we all were beginning to feel the effects of the battle. As we settled in for the night my mind was flooded with what had happened. I fell asleep almost as soon as I lay down on the ground. It seemed like I had only been asleep a few minutes when I was awakened and asked if I could give the guy something for his headache.

I looked around and saw the concertina wire that marked the darkened perimeter. Flares floated in a crude flickering ring above us casting shadowed light into a dark gloomy and forbidding world where things crawled underfoot and in the latticework of vines overhead. I soon fell into a sound exhausted sleep again.

I had agreed to stand the 0400 guard, so everyone could get as much sleep as possible. My platoon was scattered on the ground behind the bunker sleeping. It was only going to be an hour so maybe I would be able to get some more sleep when it was over. I looked to the north and saw the bright glow of parachute flares a few miles away and listened to the muffled sound of artillery as it fired in support of another unit being hit.

I leaned against the bunker and closed my eyes, knowing it was not what I was supposed to do. I bowed my head and listened to the jungle around me. I could hear the drip of water falling from the leaves and occasionally the sound of a jungle animal as it scurried through the undergrowth. I felt myself relaxing, my eyes burned, and I knew if I didn't do something soon I would fall asleep again.

I forced myself upright and took some water from my canteen to splash on my face. I needed something more than mental effort to stay alert. I remembered someone saying that the krait, a deadly snake, was a nocturnal creature. It hung from branches and would bite anything that came within reach. That's all I needed. Fear knotted in my stomach and a cold sweat made my skin crawl. I was wide awake once more. Except now I had thoughts I did not want.

I walked over to the next bunker and found I wasn't the only one having trouble staying awake. Both guys manning that position were asleep. After waking them up I returned to my position. I began to think of the day before and our four buddies we had to leave there. I thought about our leaving in a couple of hours to return to that valley to recover their bodies. We had sent twenty-three of our friends out on medevac choppers and returned with one NVA prisoner.

It was obvious that the hill was honeycombed with a tunnel system we hadn't been able to find. We had seen the enemy fall from being hit and when we got to where they should have been there was nothing but blood there.

As I looked out into the blackness, I noticed the sky beginning to pale in the east. The stars were beginning to fade as the blackness turned to gray. We would soon be leaving, and the coolness of the night was going to be replaced with the heat of the day. I was no longer tired, and I wondered what this day would bring.

Chapter Fourteen

"The truth of the matter is that you always know the right thing to do. The hard part is doing it."
— Norman Schwarzkopf

As the sun began to come up, I saw the company commander and platoon leaders gathered around the CP. The officers and platoon sergeants had been up for the past couple of hours planning our route back into the valley. Our first priority was to recover the four bodies we had been forced to leave in the valley.

It was decided we would circumvent the hill and enter the battlefield from the south while Charlie Company would go back over the trail and meet us. Gunships had begun flying over the battlefield as soon as it began to get light. The reports indicated the bodies had not been moved and no sign of the enemy was seen. They would continue watching the valley while we were on the move.

We saddled up and left the safety of the perimeter. Our patrol was to take us through a part of the jungle we had not been in yet. The maps indicated it was an easy route except for streams that meandered aimlessly throughout this region. Our nerves were on edge as we entered the jungle. The morning temperature already was close to intolerable and the sun had only been up a short time. Under the canopy the humidity hung like a visible fog. We were dripping with perspiration and going through our water way too fast.

The jungle seemed to be alive. Monkeys screeched from the tree tops and the ground seemed to be moving in places. As the guy in front of me passed under a low hanging limb I saw a pod of ants that must have numbered into the hundreds fall to the ground, just missing his back. A half hour later we passed a five-foot tan and gray colored snake laying by the trail. Someone up ahead had killed it with their machete.

Around 1100 we broke through an extremely dense part of the jungle into one of the most beautiful parts of the world I had ever seen. I was mentally intoxicated by the beauty of the vista before me. Flowing through the center of the small clearing was a stream. The jungle lushness extended all the way to the water's edge, and in places where the stream made a silent pool there were water plants flowering in the sun as it filtered through the sparse canopy overhead. Flowering bushes with blossoms like multi-colored orchids hung out over the water.

Brightly colored birds resembling parakeets flew from plant to plant, their singing lulling us into a dangerous euphoria. Indeed, a man could be at peace here. The very atmosphere had an eternal quality about it. Everyone came to a stop as we all took in the beauty. Almost as one we realized what a perfect ambush site this was and just how vulnerable we really were. We quickly moved out of the clearing and back into the jungle once more.

An hour later we came to the edge of the jungle and across the valley in front of us we could see deep craters evenly spaced like someone had designed them. These cone shaped craters were wide, and quite deep. These were the result of a recent B-52 arc light strike.

I stood looking into one of these craters with amazement as I realized the power behind the explosion which made this hole. As I stood looking into one of the craters, I remembered the freshly dug mortar pit I had stood beside the day before and wondered that something was not right about what I had been looking at. Like these B-52 craters the mortar pit did not have any dirt piled around it. These craters had been made with such force the rocks had totally disappeared. Likewise, all the reddish dirt from the pit had been carried away and was out of sight.

We soon came to the valley we had fought in and found that Charlie Company had already arrived and recovered the four bodies. They had not been mutilated and only one had his boots and weapon taken. The enemy had left in such a hurry they had been unable to collect the equipment the other three had been carrying. This was a good sign because if the enemy had not returned to gather the weapons, they were not in the area.

The tree line where the Skyraiders had dropped their cluster bombs was totally void of leaves. We could see two platforms high in the trees where snipers had stood as they fired on us. It appeared the Skyraiders had returned after we had cleared out of the valley and the fortified bunkers had taken a direct hit from a napalm canister. Plus, the five hundred pounders from the Air Force jets had done an excellent job at collapsing every bunker we found.

We patrolled around the valley for the next hour and after finding nothing we headed back over the trail we had used the day before and arrived at the fire support base just as choppers arrived with a hot meal and a bag full of mail. So much had been happening the past week that mail had slipped my mind.

We continued to patrol the same area for the next week. I stood by my bunker one morning and watched as a patrol prepared to leave the perimeter. Their objective was to go two thousand meters southwest from the fire support base. They would pass through the observation post, enter the jungle for three hundred meters, then turn south. We had been searching for a tunnel system all week but hadn't found a thing.

The patrol had just passed through the observation post when gunfire broke out. I could tell by the sound that it was an M-16. I grabbed my aid bag and ran up the hill. The firing had only lasted a few seconds, so I wasn't sure what I would find. When I arrived at recon's outpost everyone seemed very upset. Someone pointed down the trail, and I took off to see what had happened. I had only gone a few steps when I saw another medic sitting beside someone lying next to the trail.

When the patrol had passed through the outpost they had asked if anyone was out in front of them. Without checking they had been told the way was clear. The point had only gone about fifty feet when he had seen someone kneeling in the shadows. He opened up with his M-16 and had killed one of recon's guys who had gone out to check on a claymore mine.

I looked down at a new replacement that had only been in country for two weeks. I didn't even know his name. The point walked with his M-16 on fully automatic and all nineteen rounds had hit the guy's upper body, killing him instantly. I sat down and looked at the other medic who was just sitting beside the body. Our eyes met but neither one of us said a word. I looked down at my fatigue pants that were caked with dried blood from the battle three days before. Here was another needless death. I looked over at my medic friend and we both wiped a tear away.

Someone brought a body bag and we put our dead friend inside and zipped it shut.

The patrol returned to the perimeter with the body and my platoon prepared to take over. As we were about to leave, two choppers landed and out stepped two German Shepherds and their handlers. On the other Huey was a team of four guys with portable gasoline blowers. If a tunnel entrance was found, they could blow colored smoke into the hole and see where other tunnel entrances and ventilation holes were located. It was decided that the entire company would saddle up and take the patrol.

Soon we were passing through the observation post and into the jungle with the dogs leading the way. This was the first time we had patrolled with dogs and it was interesting to watch them. They were all business and at times their hair seemed to stand on end.

This time we were in luck. We hadn't gone more than five hundred meters when we found a tunnel. Shortly after we located the first entrance, we received word we were being replaced by another company. Our area of operation was being changed. We were all relieved to hear this. We had spent enough time patrolling these hills and valleys and had seen too many of our friends sent out on stretchers, or in body bags. We were ready for a change.

Choppers arrived two hours later, and we climbed aboard. Through the open door, I could see a cloud of swirling dust and debris as the chopper picked up to a hover. It tilted and spun, the sunlight flashing through the windshield and then back to the side door. The Huey began to race across the field towards the tree line. The scream of the engine increased and almost drowned out the popping of the blades. Up we popped and over the trees we raced at tree top level.

Our objective was an area the First Air Cav had patrolled

about ten months earlier and no one had been there since. We were to rebuild a fire support base they had used and be ready to receive a battery of artillery guns in two days.

When we arrived, we found an old perimeter filled with rotting sandbags and collapsed bunkers. A stream flowed about a quarter mile away and the single canopied jungle was so open the sun easily reached the ground.

Soon after we landed, another chopper arrived with an Air Force Sergeant and five large radios. He was to set up a direct relay station for calling in the air strikes. Within an hour after we landed, a Piper Cub spotter plane flew over and dropped a small canister with his mail in it, tipped its wings, and flew away.

We were rebuilding the perimeter the next morning when Sgt. Perez step out of his tent and tossed a red smoke grenade. In the distance we could hear the spotter plane approaching. Over our perimeter it flew at tree top level right towards the red smoke. As it passed a small canister fell out the rear window. Just then the left wing caught a large tree that was engulfed by the red smoke and down came the plane.

It happened so fast all we could do was stand there watching as it crashed between our newly constructed bunkers. I ran over and helped rip open the door, so we could get the two guys out. A Major who had been flying the plane was not injured but the Captain in the back seat was on the verge of shock and had a fractured wrist.

As we were getting the Captain out, the plane began to burn. All the while I had been listening to the Major swearing about some expensive night camera equipment that had been installed the night before and something about white phosphorous rockets that had been mounted under the wings.

We were laying the Captain on a stretcher when the first rocket ignited. Fortunately, it headed across a part of the perimeter we hadn't started on yet. We all got behind the bunkers and watched for the next few minutes as the rockets exploded and the expensive night camera equipment went up in smoke.

When the medevac had taken the injured Captain away, we got word the artillery was about ten minutes out. For the next four hours, we had a steady stream of choppers coming and going. On one of these choppers we received new fatigues, a hot meal, and another bag full of mail. It seemed like Christmas a month early. I had forgotten what a fresh set of clothing felt like and I was beginning to wonder if food could be heated.

With my mail was a note from Dr. Lang that said there had been a mistake. I had not made Spec. 4 as I had been told. Only Allen Gilmore had been promoted but that I was doing a good job and to keep up the good work and if I had any questions he didn't have any answers. And by the way Allen said to say, "Hi".

It was good Dr. Lang was in Base Camp and I was out on the Cambodian border. The news never made me very happy. I did feel good for Allen though. He had been assigned the job of driving for our aid station in Base Camp and if anyone deserved a promotion it was him. Working around Lt. Col. Kincade had not been easy but Allen must have been doing something right because all promotions had to go through the Lt. Colonel.

We guarded the artillery for the next ten days. We had been hearing rumors about returning to Base Camp in Pleiku for Christmas and New Year's. I hadn't seen Base Camp since September 5th.

On December 7, we packed up and flew back to Base Camp and what a surprise. All the tents had wood floors, and the roads were being prepared for paving. There was a nice PX and the camp had multiplied in size.

Just across from the aid station was an EM Club, so Corky and I headed over to get a cold coke for a dime. As we walked in we noticed a small sign that told us to take our hats off. As we stood at the bar waiting to be served we saw the bartender take a hammer and ring a brass gong. Standing next to us was a guy still wearing his hat and the house rules stated that if you came to the bar wearing your hat you had to buy everyone standing there a round of drinks.

We took our free coke and immediately found an empty table next to the bar. Every time we saw a new guy come in still wearing his hat up to the bar we went. After ten minutes, we each had a free six-pack so back to our tent we went to write letters.

Our stay at Base Camp only lasted two days. We received orders to be ready to leave in two hours. We left Pleiku City by truck and traveled east towards An Khe, home of the 1st Air Cav. Word had it we were going to guard the west edge of Mang Yang Pass. This did not mean anything significant to us except that we were to be along the highway. That meant people and traffic. Things we had not seen out on the Cambodian border.

Actually, Mang Yang Pass means more to the French than it does to the Americans. In the 1950's, several hundred French soldiers were ambushed in a truck convoy and many were killed. When we arrived at the foot of the pass we saw what appeared to be a logging clear cut up on one of the ridges. This turned out to be grave markers of French soldiers.

The highway was a two-lane black top road just like in the States. We watched as trucks and brightly painted busses loaded with people passed each day. I never realized just how much I had missed seeing civilizations until we set up our camp along this highway. We spent the next two weeks sitting around playing checkers and writing letters. I got the feeling we were sent there to get us out of Base Camp while we waited for the holidays to arrive.

About this time, we began to hear strong rumors about our group becoming mechanized after the first of the year. This sounded good to us. If we did become mechanized it would mean we would no longer have to walk but would be transported by armored personnel carriers, or APCs as they were called. To us that meant no more jungles and that was fine with all of us.

Five days before Christmas I received another note from Dr. Lang. He expressed his sympathy to me for not being promoted in November and just wanted me to know that he had put his professional reputation on the line and I was now an official Spec 4, so don't let him down. I gave him a verbal response to my good fortune but again it was best that he was in Base Camp and I was forty some miles away.

In weapons platoon were two friends, Tim Johnson and Tony Miner, who had met at Fort Lewis and began singing together. Before leaving the States, they had recorded two songs. The songs were two that Tony had written, one called "Forgotten Bells", and the other called "Anna Marie". Earlier in the month we three had been sitting around our bunker singing Christmas carols when of all people, Lt. Col. Kincade walked up and told us we would be doing a program Christmas day.

We had been practicing every spare moment we could get and had put together a good half-hour program. We were told on the 23rd that we would be flown to four artillery bases throughout the Central Highlands and that we would have two guitar players and a drummer that would go with us. This was turning out to be better than we had imagined.

We moved to a forward Base Camp west of Pleiku called Three Tango on the 24th and as usual things never quite happened as planned. On Christmas day, we all gathered around our decorated bamboo Christmas tree and did the program standing on the back of a deuce-and-a-half.

The Everly Brothers would have been jealous if they could have heard me sing their hit song, I'm A Travelin Man. Because of reasons we never quite understood, we only did the one program but to our surprise Tim and Tony received orders to join a USO show that was to tour Vietnam for the next six months. I was told that medics were not that easy to replace so I would have to stay where I was. Maybe the Everly Brothers wouldn't have been that impressed after all. I was excited for Tim and Tony but was also sad to see them go.

My mail during Christmas week held a couple of surprises. I not only received the usual Christmas greetings from my family and friends, but I received a card and letter from a girl who was a sophomore at Wesleyan University in Lincoln, Nebraska. I have no idea how she had gotten my address, but it was nice to get her letter.

The last week of 1966 went by uneventful, which was fine with me. I was tired of war and if I could have had my way I would have caught the next plane home. Out of our original group of medics, we had three that had been wounded and were in Japan. Four were in the hospital with possible malaria and would soon be gone as well.

On the 30th of December, I took my good friend Ruben to 4th Med with a high temperature. He went to sleep that night and the next thing he knew it was January 4 and he was on his way out of Vietnam with a bad case of malaria. He eventually ended up in Japan and after his bout with the malaria he was assigned to a ward at one of the hospitals there. His tour of duty was finally over.

Standing on the left with Tony Miner and Tim Johnson.
Christmas Day 1966 at Three Tango

Chapter Fifteen

"Never in the field of human conflict was so much owed by so many to so few."

—Winston Churchill

New Year's Eve turned out to be the wildest party I had ever been to. Only those with a rank of E-5 and higher could drink hard liquor but that didn't stop the rest from getting plastered. Word had come down from headquarters that there was to be no firing of weapons at midnight. At about 2350 we started to hear gunfire and moments later there was a steady sound coming from the camp. Out on the perimeter we could see tracers floating skyward resembling streams of red. It resembled Puff in reverse and it never stopped until after midnight.

Allen Gilmore and I decided we would take the watch at the aid station and about 2200 we had our first casualty. My CO from Alpha Company and two of his peers had gotten drunk and began fighting. Unfortunately, he had come out the loser. It took eight stitches in his forearm to close the wound. Throughout the night we had a steady flow of walking wounded passing through our tent. Most had fallen over tent stakes and had various cuts and bruises.

The next day, Dr. Lang asked if I would like to stay in Base Camp for a week or so and do some typing for him. I told him that was the least he could do for me after making me depressed during November and December. So, it was decided I would stick around. Secretly I wished it would turn out to be a permanent assignment. I was tired of playing warrior and even Lt. Col. Kincade didn't have the effect on me he once had.

I took a pass into Pleiku and for the first time since I arrived in country I dressed in civilian clothes. As I was walking down main street I noticed a line of GI's standing outside this small building. Curiosity got the best of me, so I walked over to see what was there. On the wall I noticed a sign that said, "Universal Auto Sales — San Francisco and Hong Kong." Beside this sign were pictures of new 1967 automobiles. It seemed you could order a new car to be delivered upon your arrival home after your tour of duty.

That sounded good to me. I spent some time looking at the pictures and fell in love with the new Corvette. For only a base price of $3,206.00, I could get a basic corvette. But then is there ever a basic Corvette? I got in line and proceeded to order myself a Goodwood Green Corvette with a 327cu. in. V-8 with a 4-speed, am/fm radio and a removable hardtop. This came to a grand total of $3,851.00. It included the $185.00 delivery charge and a promise that I could probably get about $1,400.00 trade in for my 1964 Nova Super Sport.

What more could I ask for. I put $100.00 down and agreed to pay $60.00 a month until my tour was over. The monthly payments would be about $75.00, I was told. The delivery date was to be August 1st and the dealership in Seattle was picked. Now all I had to do was pay my $60.00 a month and go to sleep at night dreaming of my new Goodwood Green Corvette with a 327cu. in. V-8 with a 4-speed, am/fm radio and a removable hardtop. I only stayed a couple of hours in Pleiku and spent the rest of the day at Camp Holloway, a helicopter base. I had a steam bath and massage and when it was over I noticed the red dirt that had been a part of me for the past four months was gone. No matter how hard I had washed, it had not gotten out of my skin until now.

The battalion moved to Three Tango and word was that we were in fact going mechanized. We also heard rumors that when we had our APCs we would be transferred to the Delta Region below Saigon.

On January 7 Dr. Lang called me into his tent and told me I had an hour to catch a chopper heading for Three Tango. I knew my temporary job as typist would not last forever, but I wanted it to last at least another month or so. I promised him I would take typing by correspondence and would do his sick call every Monday, Wednesday, and Friday, but he said the war was just not the same without me, so have a nice flight.

Soon we left the populated city of Base Camp and headed into open country. We flew over a hill top where a group of engineers were working with their bulldozers. The green of the jungle had been stripped away to reveal the reddish earth. Puddles left from that morning's rains winked up at us in the afternoon sun.

When I arrived at the forward aid station, I found I would not be going back to my unit for a while. I was to stay there and help. Two nights before my arrival around twenty-five mortar rounds had landed about a thousand feet from the aid station. Eleven guys had been injured.

As I was putting my things away I saw what appeared to be a small dog coming out of Bravo Companies supply tent. I watched as it headed my way and to my surprise it was not a dog but a baby sun bear. One of Bravo Companies patrols had killed its mother when she charged their point and then they found the cub. It was to be their mascot and they had named it Bravo.

On the 9th I was told to get my things ready. I was to go replace a medic from Charlie Company who was needed in Base Camp for something and I was to only be gone for a few days. Charlie Company was guarding a group of engineers while they built a road through the jungle about fifteen miles northwest of the Special Forces camp at Plei Djereng.

I packed a rucksack for a short stay and headed for the heliport. When I got there, I found a Huey waiting and as I climbed aboard the pilot began to wind the helicopter up to full operating RPM. The noise increased as the turbans began to roar, and the blades popped as we lifted off the ground. The pilot dropped the nose and began to fly east away from the perimeter. As we approached the tree line he pulled up on the cyclic and popped over the trees. He stayed there flying ten feet above the tops of the trees for a mile or so before climbing to the cruising altitude of 1,500 feet.

I had mixed feelings about going back out into the bush. It had been a month since I had slept in the jungle and I had not missed it once. When I got off the chopper I found a real logging operation going on. Except this time, the trees were not the firs and pines found in the Pacific Northwest, but the teak and mahogany and other exotic trees found in Southeast Asia. The engineers were making a road to the top of a large mountain where a fire support base was to be placed. They were making what appeared to be a two-lane road with a field of fire on either side of a hundred feet or so.

Soon after I got there I asked where the toilet was and was pointed towards a path and told I couldn't miss it. Around the bend in the trail I saw a fifty-five-gallon drum that had been cut in half and a seat made for it sitting next to a pile of trash. As I was returning to the camp I passed another guy heading for the john. He was carrying his machete and as we passed he said something about needing to sharpen it.

Five minutes later he returned carrying a six-foot cobra. He was sitting on the john and happened to look up into the beady little eyes of this cobra. The snake was almost between his knees and had raised up a couple of feet and was looking at him with its hood wide open.

This had a strange effect on me, to say the least. At least this guy had his machete with him and was able to hit the snake before it got him. Little cold beads of sweat trickled down my back and I felt my knees getting weak. It was the first cobra I had seen. I knew they were there because others had killed them, but I had never seen one.

To my pleasant surprise, I was recalled to Base Camp after staying two days with Charlie Company. Dr. Lang's note said something about needing a typist that was slow enough the typewriter could not be damaged. I was to up-date shot records and prepare a couple hundred personnel records. I could see at least a month's work if I did it right. As my replacement stepped off the chopper, I got on and headed towards Base Camp and a shower.

The one thing I noticed when I arrived at Base Camp was the difference in temperature. There was at least a ten to fifteen degree spread between the jungle and Dragon Mountain Base Camp. The mornings at camp were hitting in the high 40's and the wind seemed to blow all the time, but I was just glad to be there.

I finished my job at Base Camp on January 23rd and was actually glad to get back to Three Tango. It was cool in Pleiku and even though we were only twenty-five miles or so west the warmth felt good. I was scheduled to go on R&R in a couple of weeks, so I would stay and work at the aid station there.

The next morning, I was informed I would be part of the med-cap unit that would be leaving in an hour or so. I had never heard of a med-cap unit and was told it was a medical group and intelligence group that went out to the surrounding villages and did sick call after the Intel people got the information they needed.

After an hour ride we rounded a corner and up ahead I could see a village made of bamboo huts with thatched roofs. One or two were roofed with corrugated tin. I could see an occasional ox cart and the oxen were tied to poles or in pens made of thorn bushes and logs. The streets were hard packed dirt paths that meandered through the village.

The huts which lined these paths were a mixture of old and new, all made from bamboo and thatch. Smoke from cooking fires hung low to the ground. Behind the huts, I could see smaller ones built on stilts about four feet off the ground. I was told these were packed to the roof with rice. And everywhere were children.

We stopped the ambulance in the village square and the Intel people went to work asking the headman of the village about enemy movement in and around his village. I noticed that the children and most of the adults had all gone to the far side of the village. I could see what appeared to be a pond of some kind and they were bathing. These people knew they had to be clean before they could visit the doctors. I walked over to the bathing area and handed out new bars of soap to those who wanted them.

Soon they came for treatment and for the next hour we cleaned open sores, drained abscesses and boils, and dressed them with sterile bandages. Infected scalps were scrubbed with surgical soap, and instructions given to bathe every day. The sad part was seeing those who were terminal with TB and other diseases. All that could be done for these people was to give them something for pain.

It was sad to see these people living in such squalor. I had never seen anything like it. To see the children and realize they would probably never know anything better. Most of these people wore the native loin cloth and covering their naked bodies was sores filled with infection. It was obvious they were glad we had come. Almost the entire village stood in line to see the doctors, even if it was just to get an aspirin. One of the first in line was an old man who said he had a cough. We gave him a Cepacal cough drop and the look of delight on his face told me he thought it was candy. Pretty soon we had others in line who said they too had a bad cough and needed one of the yellow pills. We probably gave out more Cepacal that day than any other form of medication.

On the 26th, Charlie Company received a radio call from someone requesting transportation for two medics that were waiting at the fork in the road about two miles away. The CO called the senior aid man over and asked if all the medics were accounted for. The NVA had found out Charlie Company's radio frequency and were attempting to lure them into an ambush. Instead of a jeep showing up at the fork in the road the 155mm artillery guns used it as a target.

On February 1st, I was told I would be leaving for R&R on

the 21st of the month. I was also told I would be going back to Alpha Company. It had been quite a while since I had been with them. Knowing I had to go made it easier to take. By the time the chopper took off I was glad to be getting back to my friends and when I landed it was obvious they were glad to see me as well. Even Lt. Underwood came up and said he had missed me.

I was told that Alpha Company had just arrived at the fire support base and we would be pulling guard duty for the next week or so. The fire support base was a new one. It was located at the end of the road the engineers had been building a month earlier. It sat high on a mountain overlooking the Cambodian border and was loaded with fire power. It had five 105mm, three mounted 155mm and two mounted 8" guns. The top of the ridge had been cleared. The cleared hilltop was made of white sandstone. It was so bright at night we were able to write letters by moonlight.

At Dragon Mountain Base Camp, January 1967

Frank Corky Colburn with Bravo the sun bear cub

Chapter Sixteen

"Whoever said the pen is mightier than the sword obviously never encountered automatic weapons."
—Douglas MacArthur

The new fire support base was like none I had ever seen. It must have been something special because the next day General Westmoreland came to inspect it. He said it was the best example of a fire support base he had seen. Our perimeter was ringed with three rows of concertina wire, and barrels of foo-gas. Machine guns manned every strategic point and the mountain top commanded the surrounding valleys on three sides. The Cambodian border was not far beyond the valleys to the west.

We had been there for three days when I was told I would be going on a seven-man patrol. We were to go two thousand meters out and then circle the base. We were to spend the day just looking around. At this point in my tour the idea of a small patrol didn't excite me in the least. I wasn't even fond of platoon size patrols anymore. Something about combat had taken the excitement away from this kind of adventure.

The plan was to ride on the back of a truck and as it went around a corner we would jump off and melt into the jungle like fog without being seen. As we drove out of the perimeter I secretly wished that someone would break an ankle when we jumped off the truck. I didn't even mind if it would be me.

At the appropriate point, we jumped and melted into the jungle like fog. Two hours later we had finally calmed our nerves a little, so we took a break on the side of a hill to study the map. Jerry Patterson sat down beside me and said he hadn't been feeling very well. I looked into his eyes and clearly saw someone who was coming down with malaria.

I stuck a thermometer in his mouth and to my surprise he was running a temp of 104*. I gave him some aspirin and had him drink as much water as he could handle. I went over and talked to the patrol leader and we searched the map for a stream. Finding nothing close by, we considered calling a dust-off for him.

We reported our possible need for a dust-off and asked for a spotter round to make sure we knew just where we were. As the round popped above us, we all froze in place. Just a hundred feet below us on what appeared to be a trail we had seen movement. We silently spread out into a cautious unmoving crouch behind the nearest tree with weapons pointed all around. We peered through a gap in the brush, our attention focused on movement, black movement, at the bottom of the ravine.

Because of Jerry's high temperature he was beginning to appear as if he was drunk. About this time, he decided it was time to party. He started laughing and told me how funny it would be if we got caught. Way to go Jerry, only seven of us in the middle of a jungle and over a mile from anyone friendly. I quieted him down and took his M-16 and grenades away from him just in case he decided things were funny enough to disclose our hiding place.

Staying motionless next to the giant teak tree and without a breeze, the sweat began to trickle down my face and back. I looked at Jerry who was beginning to become lethargic. I was worried because his temperature was so high and the only thing I could do was give him aspirin and water until we could move up the hill and get him medevaced out.

We watched in stunned silence for five minutes as the black movement continued below us. Just as our hearts were about to stop, a large wild pig walked into view. Seven M-16's took aim at this monster but we all knew better than to fire.

We had been in one place too long and had made a fair amount of noise. Jerry was now in a state of inertia and his temperature was holding around 103 degrees. He needed help soon. We decided to call the patrol off and get him to a hospital. Within the hour, he was headed for 4th Med in Base Camp and within a week we got word he had tested positive and was in Cam Ranh Bay.

As we came into the perimeter another patrol was calling saying they were about five hundred meters out and would be coming into view soon. Their RTO had just finished talking when we heard gunfire coming from their direction. We could distinctly hear both M-16's and AK-47's. After a minute of long, sustained firing, we were able to contact them again. The point had surprised a squad of NVA building sniper towers. This was a good find for us being as the towers had a clear view of our entire perimeter.

That evening we were told our APCs were arriving in Qui Nhon. Rumors had been strong that we were going mechanized, but this was the first positive indication that the rumors were in deed factual. I was now assigned to the recon platoon and we were told that our platoon would be the first to get the new tracks. The rest of the battalion would have to wait another month or so.

The next morning second platoon returned to the sniper towers to set up an ambush and it turned out to be worthwhile. They had waited about ninety minutes when another squad of NVA came into view. Second platoon killed one and was able to capture another. He had been wounded four times but was in no danger of dying. He told us that an NVA regiment was planning on attacking the fire support base in a couple of days. This was not what we wanted to hear but, like a lot of information taken from prisoners, the chances were just as great it wouldn't happen.

The amount of contacts our units had been making since the first of February was an indication of another build-up in our area of operation. That night two Air Force jets, returning from a bombing mission, spotted a lot of lights crossing the river on the border. Around midnight we watched as B-52s saturated the crossing area. Their display reminded me of WWII movies I had seen.

A company from another battalion was patrolling the valleys directly to our west and had been watching undetected for two days as a large number of NVA had been gathering on one particular hill. This hill was clearly seen from our vantage point. It was decided that the rest of their battalion would make a combat assault the next day.

As the sun came up, jets began attacking the hill with five hundred pounders. We could see the bombs fall as they dove one after the other. When they finished bombing, our artillery began to blanket the hill and soon the sky was filled with Hueys and gunships as the battalion headed into the LZ. As the battle progressed, we saw five Hueys heading our way. Without warning, all five made emergency landings inside our perimeter. All of them had taken many AK-47 rounds as they had left the LZ. Two of them had to be carried out by helicopter cranes.

Before the day was over, the planned attack on our fire support base had been taken care of. The NVA took heavy causalities and headed back across the border. On top of the hill in a fortified bunker was found a dead Chinese officer and documents saying that our fire support base must be destroyed at all cost.

It was February 8th and as far as I was concerned, the NVA needed to wait until after I was gone on R&R before they came back from their camps in Cambodia. Earlier that week Harold Stenseth and Dr. Lang had headed to Hawaii for their R&R. Both were meeting their wives and would spend five days enjoying the islands. I envied them as I lay on top of my bunker that night and looked at the stars.

An interesting story was unfolding and would not be recognized for its real meaning until years later. Harold had been on patrol with his platoon when he received a radio message late in the afternoon. He was told to catch the next chopper in to Base Camp and report to the Cam Ranh Bay R&R center the next day. He had not been scheduled for R&R for another couple of months.

Two things concerned him. His wife Carol was unaware of this change and the other concern was financial. Being between pay periods his cash flow was virtually non-existent. By the time he received this message, it was on the verge of getting dark. He was told to get ready because the last chopper was already inbound.

When he arrived in Base Camp, the first person he saw was Allen Gilmore. In their conversation, Harold mentioned that he needed some money. Allen took him into his tent and opened his foot locker. He took out an ammo box and opened it and to Harold's surprise it was filled with money. Allen had been saving his payroll each month and had saved quite a large amount. He began counting out the bills into one hundred-dollar stacks.

For those of us who knew Allen, this did not seem unreasonable. A simple hand shake was all that was needed for this loan and one of Harold's two concerns was taken care of. He then went to the MARS radio station and within minutes was talking to Carol. Of course, she would meet him in Hawaii so the next day they both boarded planes and headed for their honeymoon vacation in the islands.

While they walked the beaches and hills of Hawaii they renewed their wedding vows they had made just a couple of weeks prior to our leaving Ft. Lewis. With emotions tugging at them, they came to the fifth and final day together. And wouldn't you know it Harold came down with malaria. He entered Trippler Army Hospital and Carol made the decision to stay in Hawaii.

Meanwhile things on the Cambodian border were beginning to heat up once more. The Lunar New Year truce ended, and it became evident the enemy had used that time to prepare for battle. On the night of the 14th they attacked Three Tango, Plei Djereng, and units from the 1st Battalion 12th Infantry and 1st Battalion 22nd Infantry. The road between Three Tango and our fire support base had been heavily mined.

On the morning of the 15th, a jeep hit one of these mines, seriously injuring the three guys riding in it. Corky's platoon headed out to assist and came under small arms fire. During the short firefight that ensued, they had two wounded. When they returned to the fire base, the Lieutenant looked like he had been knocked down and run over. Just as he bent down to pick something up a bullet had hit his helmet. Besides knocking him flat, the bullet pierced the helmet and spun around the inside finally stopping and dropping out onto his shoulder. This amazing bit of luck had only given him a tremendous headache and put burn marks around his scalp where the bullet had spun.

Around 1400 on the 15th, we got word that a platoon from Charlie Company was being hit. Their point had walked into an NVA Base Camp without knowing it. The platoon had turned and began running towards higher ground but before they could reach it they were surrounded and pinned down. Alpha Company was told to be ready to re-enforce them. We saddled up and waited for choppers to get there.

At 1630, we were told to get to a larger LZ about four miles away. We had begun receiving some of our new APCs a couple of days before, so it was decided to have them transport us. We climbed on and headed towards the new LZ. The sun was getting low and we had never made a night combat assault before.

Just as the last rays were sinking in the west, the Hueys arrived. The number of choppers and gunships indicated this was a serious matter. The choppers came into the LZ in groups of seven, and we climbed aboard. After a thirty-minute ride, we began to circle. Below us Puff was dropping parachute flares, and from my seat on the side of the cargo bay I could see dense jungle and what appeared to be a point of land the choppers were heading for.

As I watched it began to dawn on me where we were. This was the same finger of land by the deserted village where we had spent the night with Charlie Company back in November. The reason for our having to circle was that only one chopper could land at a time. They were touching down on the log platform Charlie Company and the two engineers had built.

Corky and I were on the seventh chopper to land and the pilots and door gunners were nervous. Corky was behind me and was pushed out of the chopper when it lifted off and headed out of the valley. We had barely cleared the platform before another chopper landed. We were fortunate that the enemy was not around. It would have been a disaster if they had been.

Once the whole company was on the ground we were told that communications had been lost with the platoon and we only had their last grid coordinate to go by. It was decided that recon platoon would stay behind and secure the finger of land just in case the rest of the company needed to return to a safe perimeter. The rest of the company headed out through the village to find the lost platoon.

Puff had only stayed around long enough to give us light to land. As the darkness closed in around us, we wondered what would happen before morning. Recon platoon was not up to full strength and our small group seemed strangely out of place as we searched into the darkness around us. Any moment we expected to hear gunfire from the rest of the company as they too walked into an ambush.

No one slept as we waited for the call to connect with the rest of our company. Sgt. Moran crawled over to where I was and pointed towards the west where the ferocity of Puff's miniguns flared like summer lightning among the massive teak trees. I thought about the timing of this adventure. I was scheduled to leave on the 17th for my R&R. Why couldn't this have happened a couple of days later. By then I would have been in Cam Ranh Bay preparing to fly to Hong Kong.

As the darkness turned to grey, we received word that the missing platoon had been located and we were to link up with them as soon as possible. We were soon deep in the jungle and were all extremely nervous. Around us were signs of the enemy's presence. Hammocks still strung between trees and enemy backpacks lying under them. Rice bowls were sitting beside fires that had gone out. We expected to be ambushed at any moment.

At times, we were running flat out in an attempt to reach the rest of the company. After an hour and a half, we arrived at the battlefield. We weren't prepared for what we saw. As I walked into the perimeter, Corky came up and told me that most of the platoon had been killed. He had spent the night crawling around the perimeter checking for anyone who may still be alive. Mixed with the dead platoon were NVA bodies. I asked who the medic was, and he told me Joe Foran and that he was dead. I asked him where he was, and he pointed to a row of bodies covered with bloody ponchos.

It was then I realized that the platoon had been Ruben's old platoon. If he had not gotten malaria he would have been with them and would have been killed. When our company reached them around 0400, they found only three guys who could move. Only one had not been wounded and he was down to six bullets. Lying around them were dozens of enemy bodies.

The official write up in the Stars and Stripes said there were eighty-four communist bodies found. This did not include those bodies found the next day as the surrounding area was patrolled. It soon became clear that the twenty-eight-man platoon had been hit by an NVA battalion which was part of the 66th NVA regiment.

Our job of getting the wounded evacuated was not going to

be an easy one. The jungle was very thick, and the canopy was about a hundred and fifty feet above ground. A chain saw was lowered to us and we went about cutting a clearing. A Chinook arrived and lowered itself down into the small opening until its blades began to snap at the canopy. One by one we lifted the wounded in a wire basket to the safety of the chopper.

It was a slow process and the choppers could not stay there long before they needed to return to their base to refuel. We worked as quickly as we could and soon had all the wounded out except one. He had been hit in the right knee and could not be lifted out in the basket. We would have to figure out another method of evacuation for him.

As the morning wore on, we lifted the bodies out and by noon the only one left was the guy with the injured leg. Even though he was doped up on morphine the pain could clearly be seen in his eyes. We watched as a small bubble chopper hovered over the opening and began to descend. There was barely enough room for it to clear the trees, but it soon was on the ground. We strapped the guy in the basket and the pilot lifted off. We held our breath as he slowly climbed higher and higher until he reached the top and was gone.

There had been a command chopper flying overhead most of the morning and we received word the Lt. Colonel wanted to land if it was possible. We cut a few more trees down and soon a Huey was sliding down the opening. As I watched it land, I realized this was probably going to be my last ticket out for some time. I reminded the Captain about my R&R date and he gave me permission to leave.

A feeling of relief flooded through me as we reached the top of the canopy and headed east towards Base Camp. What I had just witnessed was going to change many families for the rest of their lives. Over twenty of our friends had died which meant over twenty families would soon be getting a visit telling of the death of their son.

Death! You get immune to seeing it, you learn to control your emotions, but it never quite goes away. You see the broken bodies and you smell the odor of death. It no longer causes physical sickness, but it is always there to remind you of what war is about. Vietnam was nothing more than a man trying to survive and if we did, then we would win our own personal war. No one could do better than that.

Chapter Seventeen

"Out of every one hundred men, ten shouldn't even be there, eighty are just targets, nine are the real fighters, and we are lucky to have them, for they make the battle. Ah, but the one, one is a warrior, and he will bring the others back."

— Heraclitus

On February 19th, I sat in the air-conditioned EM club at the R&R center in Cam Ranh Bay, waiting for my scheduled flight to Hong Kong the next day. I couldn't help but recognize the timing of my R&R. The enemy was clearly back on our side of the border and seemed ready to fight. I was more than willing to let them do their thing as long as I was not there to take part in it. Coming to the rescue of Ruben's old platoon, and seeing them dead, had left an imprint on my memory that was to never quite go away.

Over the past month and a half, I had sent eleven of my company out with malaria. While I waited for my flight I headed over to visit my friends who were there recuperating. The hospital was located right next to the ocean and its white sandy beaches. The first one I found was Jerry Patterson. He was walking along the beach in his blue hospital pants, and he looked a lot better than he did the last time we had seen each other. The only good thing about getting malaria was the two month's recuperation period after the week or so of being sick.

The next day I boarded a prop-jet and headed out of Vietnam. It was evening when we landed and were taken to the processing center. On the wall was a board with brochures describing each hotel. There was one brochure for each available room. The hotels gave special rates for those of us on R&R, and for only $7.50 a day I could stay at the Hong Kong Hilton, but when it was my turn to pick I found that all the available rooms there were taken, so I settled for a $6.25 a day room at the Grand Hotel on the Kowloon side.

I checked into my room and then headed out to see what was around the hotel. Most of the shops were closed, but the evening was warm, so I walked for about an hour. I was returning to the hotel when someone lit a string of firecrackers. As they began to explode I immediately stepped into a doorway and then realized I was not in Vietnam. I felt embarrassed and hoped no one had been watching.

It had been over seven months since I had slept in a regular bed, and between clean white sheets. I turned on the television and began watching a Chinese play. I couldn't understand a word they were saying, but it was not hard to follow the plot. Broken relationships are the same in any language. The next thing I knew it was 0300 and the TV was still on. I turned it off and fell into another deep sleep.

I spent my five days wandering around Hong Kong, and all too fast I came to my last night. I was determined to eat in style, so I decided to find the nicest dining room I could and eat whatever I could afford. I took the ferry to the mainland and walked into the Hong Kong Hilton. I was seated in this gorgeous dining room with pillars and crystal chandeliers. It was alive with soft music and soft lights.

Sitting across from me was a European family with a distinctive British accent, quietly talking and enjoying their meal. I looked at the menu and chose a dinner that sounded delicious. After finishing the meal, I traveled back across the bay to my hotel.

I was up at 0400 so I could enjoy my last luxury for another six months. I soaked in a tub of hot water for an hour. I had an early flight and when the ride to the airport arrived on time I was prepared to return to whatever awaited me.

When I got back to Cam Ranh Bay I learned my flight to Pleiku wasn't leaving until the next day. I headed over to the EM club to write letters and get a cold coke.

I had just started a letter to my folks when I recognized a guy from Bravo Company sitting across the room. It looked like he was alone, so I walked over and sat down beside him. He had just arrived and was heading for Bangkok the next day. We talked about the difference between being on the Cambodian border and being on the coast, and then he asked if I had heard anything about the fighting that had taken place while I had been gone.

He said that on February 21st Bravo Company had been on patrol and was late getting settled for the night. They found a small hill and sent out several patrols to see what was around them and set up a perimeter just before dark. What they did not realize was that a group of NVA had reached the hill just before them. When they had seen Bravo Company coming they had climbed the trees, expecting them to keep going, but when they settled in for the night the enemy was trapped.

During the night they began dropping grenades and Bravo

Company started picking up wounded. It was some time before they realized there was enemy inside their perimeter. As morning came, they also began getting hit with mortars.

Art Collins was tucked away in the fold of a teak tree when he saw a guy across from him get hit by a sniper. He could tell the direction the bullet had come from, so he peered around the tree he was hiding in and began searching. His eyes focused on the sniper and he realized they were looking at each other and the sniper rifle was aiming at him. Just as the sniper pulled the trigger Art jerked back into his hiding place and slumped with a chest wound.

If he had not seen the sniper first the bullet would have gone through his heart. As it was it entered his chest on one side of his heart and exited his chest on the other side of his heart without penetrating deep enough to kill him.

When the battle ended, one of the five medics had been killed and the other four were wounded. I sat there too stunned to say anything. Finally, I asked him who had been killed. He wasn't sure what the name was, but it was the one who had taken Harold's place. He described him as being tall, blonde, wearing glasses, and having a southern accent. I asked if the name Allen Gilmore sounded familiar and he said yes.

I sat there feeling like a truck had just run over me. I could feel warm tears welling in my eyes and beginning to run down my face. My friend from Bravo Company sat there, not quite sure what to do next. I turned and looked out the window. I didn't care who saw my tears. It wasn't the first time we had seen guys crying over the loss of a friend, and it wouldn't be the last time. I got up and said something about taking a walk and headed out the door.

I was standing outside the building we were sleeping in when some guy from the 1st Air Cav walked up and asked if I was all right. I told him about Allen's death, and I saw tears begin to fill his eyes. He told me his best friend had been killed a month earlier, so he knew how I was feeling.

This was the battle that also wounded several others from our group who would spend weekends at my folk's place. Gary Booth was severely wounded in his leg and when I saw him a few months after my return home he wore a metal brace.

Dio Rader's platoon was one of the patrols sent out to check around the hill. Not far from the perimeter they found a well-used trail. While on this trail they began receiving small arms fire and he heard a call for a medic. After dressing the guy's wounds, he was able to get him back down the trail a little way. He returned towards the front of the column and found two others, but they were both dead.

Their CO ordered them back to the perimeter. Once they had reconnected with the rest of the company he was treating another wounded when he was hit by several pieces of shrapnel in his face. The wounds were not deep, and he was able to stop the bleeding. As he was crawling to assist another wounded he was hit again, this time in the hip. This wound caused some difficulty moving around but at this time he and another wounded medic were the only two functioning corpsmen in the company. When the battle finally ended, there were sixty-eight WIA and six KIA. Of the six medics in the company, one was KIA and the other five were WIA. For his bravery, he was awarded the Bronze Star.

Lt. Hunter was killed by a sniper in this battle. Bravo and Charlie Companies were awarded a Presidential Unit Citation for their gallantry during the battles beginning on February 16 and ending with this battle on February 21, 1967. Bravo and

Charley Companies suffered 169 KIA and it was reported that the NVA suffered 733 casualties.

The next morning, I caught a ride to the airport for my return to Base Camp. I stood under the tail of the C-130 in an attempt to find shade. What a contrast from the air-conditioned terminal across the runway. Here on the tarmac the difference was startling. I wiped my forehead on the back of my hand. In the terminal, it had been cool enough to be uncomfortable, and here under the tail it was hot enough to be miserable.

When I arrived at Base Camp I was told to catch a chopper to Three Tango. I headed for the heliport and found the Huey sitting on the tarmac, its turbans whining quietly and the blades slashing through the air with an inaudible swish. I ducked and ran under the blades and climbed into the side well of the cargo compartment. The door gunner climbed in behind his M-60, and the turbans increased their whine. We soon lifted off, hovered a moment then turned, climbing out over the perimeter wire and red dust.

I looked out of the cargo compartment door. Below were farms with conical capped men, behind water buffalo, plodding down uneven rows. At the edge of the trees I could see thatched hootches filled with lice and spiders. The farm land was giving way to forest, which soon gave way to dense jungle, with a double and at times a triple canopied ceiling. I looked over at the door gunner and started to say something then realized words were lost in the screaming of the turbans, and the popping of blades. Sound reverberated through the chopper, making it impossible to hear.

The only other passenger was an old Montagnard man carrying a small pig with its legs tied. I sat there and thought about my last week. A week spent on jet planes, and in foreign countries. I wondered if the old Montagnard had ever been more than a few miles from his birth place.

I had mixed feelings about going back to the field. The newness of Vietnam was gone and had been replaced with a mature desire to stay healthy. The short time I had spent at Base Camp was long enough to see the hurts being felt by all over Allen's death. I heard that Allen had been taking care of another medic who was wounded, and when they began receiving mortars he was not able to move him. Instead of getting protection for himself he had laid over his friend and a round had exploded beside them. Allen had taken the full force in his back, but the other guy was protected from further injury.

We landed at Plei Djereng to let the old man and his pig off, then went on to Three Tango. Upon my arrival, I found I was to stay there for a couple of days and help the new doctor. Our replacement for Dr. Lang had arrived while I was gone. Dr. Sallel was more serious than Dr. Lang, but just as friendly.

Two days later I was assigned to recon platoon. This was great news. Recon was the only unit that was already mechanized. Their APCs had been delivered around the 10th of February, and my new home was to be the large fire support base I had been at just prior to leaving for Hong Kong. When I checked in with Lt. Campolo he recognized me from the maneuvers back at Ft. Lewis when he asked me the "what if" question. He told me he felt safer just knowing I was around.

Our duty was quite simple. Every morning we would be the first ones over the road to make sure there was not an ambush or mines waiting for the unprotected trucks and jeeps that would travel between the fire base and Three Tango. Throughout the day, we would position our APCs between the two points, or just stay around the fire base to be used as needed. We would also be the last ones to travel the road each day. The last track in our patrol would pull a log behind it to clean the road of any marks. That way in the morning we could see if anyone had crossed over during the night, and it would make it harder for the NVA to lay mines.

On my first morning patrol, we were on a stretch of road that had banks that were eye level with us. I traveled with the Lieutenant and platoon sergeant, and I was standing in the open hatch behind the LT who was manning the 50-cal machine gun. We were always watching for an ambush, and I was looking for anything which looked out of place.

All at once my eyes locked in on what looked like a twisted limb lying next to a stump. I followed it up about two feet and right into the beady eyes of a cobra with its hood spread. He was coiled and watching us pass and was maybe fifteen feet from me.

Every Eric Hare story my mother had read to me when I was little came passing before me. Especially one about how the cobra would open its hood and hypnotize small animals, so they couldn't run away, then it would strike. I felt my legs getting weak and I slumped to the seat below me. The platoon sergeant grabbed my arm, thinking I had been hit. It took over a week for the ribbing to stop. I mean if a little snake could do that to their Doc what would something bigger do to him.

When we made the first patrol each morning we were allowed to recon by fire. This meant we could fire the weapons at possible ambush sites. This might spring an ambush before we entered it. On our track we had an extra M-79 grenade launcher, and M-16. I began firing these weapons during these morning patrols, and at first this surprised the guys in my track. They thought I was opposed to even firing a weapon and were quite surprised that I knew how to use them and was actually a pretty good shot. It seemed innocent, and I was unaware of the effect it was having on the Lieutenant.

On the 3rd of March, we had returned from our morning patrol and were laying around waiting for something to happen. About 1030, the large 8" guns began a fire mission. When these guns fired, the hill would shake, and dust would rise.

I was laying on the seat of my track when there was an explosion. I sat up and asked the guys around me what had happened. We all looked around and seeing nothing unusual they told me I had been in country too long. I agreed with them and laid back down, but I had heard something different about the sound of the gun firing. It was like the difference between an M-16 and an AK-47. Each had its own sound.

I picked up my aid bag and began walking up the hill. I was about halfway to the top when I saw two other medics running towards a bunker on the perimeter. I then saw two wounded staggering towards me from the CP bunker. One of the 8" guns had had a muzzle burst. This is when the shell explodes as it leaves the barrel, and the round had exploded directly over our perimeter.

When I reached the two from the CP bunker three more were being brought to me. They had all been wounded with pieces of shrapnel. I was told that there was someone from the artillery group with a serious leg wound. When I had taken care of the five wounded I called a dust-off.

We were standing by the CP bunker when we heard moaning coming from the other side. I ran around the bunker and there was a guy laying in a pool of blood. I could see that a portion of his head was gone. I knelt beside him and he gave a sigh and stopped breathing. I told someone to call another dust-off and to request that a doctor be on board, and to go find another medic to help me. I then began clearing his airway. His mouth was filled with sand and brain matter. Spinal fluid was seeping from his ears and nose. As I started working on him one of the infantry guys began to do artificial respiration. Soon two more medics showed up and we were able to get an IV going. He had lost a lot of blood and was not able to breath on his own.

In fifteen minutes, the chopper landed and Dr. Sallel got off and came over to where we were. Someone lifted the large bandage that covered the head wound so he could see the extent of the injuries. It was obvious this was the worst injury Dr. Sallel had seen so far in his short time in country.

As we were loading the wounded on the chopper, I asked if he wanted me to come along to help, and he told me it was no use. The guy was going to die anyway, there was too much brain damage. He was surprised we had been able to keep him breathing at all.

We soon got word that he had died enroute to 18th Surgical Hospital in Pleiku. His death set the tone for the rest of the day. He had been another new replacement that had been in country for only a short time and was the victim of another accident. It showed us once more just how fragile we really were.

For the remainder of the day things were unusually quiet. Most everyone kept to themselves, or talked quietly in small groups, but by evening we returned to normal. We were used to death by this time and had a built-in safety for our emotions. We knew it was not a good idea to dwell on the violence for very long.

Our daily routine never varied much. We would usually position a couple of APCs' along the road, but the one I was on would return to the fire base with an occasional patrol along the road throughout the day. Whenever there was a convoy coming from Three Tango, we would join it and give security.

A couple of days after the artillery accident, my track was sent about four miles down the road to check on an accident. When we arrived, we found that a five-hundred-gallon rubber container filled with aviation fuel had come loose from the back of a deuce-and-a-half. It had rolled down an embankment about a hundred and fifty feet and was lodged under a fallen tree.

We radioed back to the fire base and told the Lieutenant that it couldn't be salvaged and were told to destroy it. Everybody took their M-16's and began shooting at it. I was using the Lieutenants new CAR-15, which was a modified M-16 that had a collapsible stock. We had fired a couple of hundred rounds, including tracers into it and it still had not exploded. I climbed into the track and got the grenade launcher with a regular grenade and a white phosphorus grenade. The track driver was standing beside me when I fired the regular grenade. It hit a small tree beside the container and when it exploded the driver swore and stepped away from me. I thought he was mad that I had not hit the fuel. I then loaded it with a WP grenade and this time the fuel went up in a ball of flame and cheers from all of us.

When we returned to base, I noticed a spot of blood on the track driver's pants. I told him to drop his pants, so I could look. There on his hip was a tiny cut. I told him he would have to go back to Three Tango for a tetanus shot. I assured him it was a minor injury and he could come back in the afternoon.

Three days later, he returned and immediately came looking for me. It seems the cut was caused by shrapnel from an M-79 grenade. The doctor had probed around in it and had found a tiny piece of shrapnel. I don't think I ever lived that one down. I became the Conscientious Objector who wouldn't kill the enemy but had no objection of wounding his own buddies.

Dr. Sellel at The Oasis June 1967

Chapter Eighteen

"...It is a proud privilege to be a soldier—a good soldier... [with] discipline, self-respect, pride in his unit and his country, a high sense of duty and obligation to comrades and to his superiors, and a self-confidence born of demonstrated ability."
—George S. Patton Jr.

One afternoon in late March, we received word that one of our tracks had hit a mine. It had been leading a convoy from Three Tango and was about five miles out. This was unusual because it meant the enemy had planted the mine during the day. We had cleared the road that morning and had found nothing.

We loaded up and headed for the convoy. When we arrived, we stopped a quarter of a mile away and began sweeping the road with mine detectors. We soon found another one and dug it up. Usually the NVA planted three mines at a time, but after looking for a half hour and finding nothing else we decided to let the convoy continue.

It was then we found the other mine. A deuce-and-a-half had stopped about an inch away from hitting it. It had been so close to the tire the mine detectors had not found it. The truck was pulling a water trailer, and as it began moving I was watching a guy walking beside the trailer. The explosion picked the front of the truck off the ground, and I saw what looked like a body fly into the air about a hundred feet. The result was electrifying.

I jumped from the APC and ran towards the truck. I could see someone lying about forty feet off the road, and I was sure I would find him without legs and probably dead. When I got there, it appeared he had laid down and taken a nap. Even his arms were folded across his chest, and his helmet was still on his head. Lying about ten feet away was the shredded tire from the deuce-and-a-half. That was what I had seen fly into the air.

I checked his vital signs and they were pretty good considering, so I felt around for broken bones. The guy still had not come to, so I took my canteen and poured water on his face and he woke up. He wasn't sure where he was and said something about not being able to hear very well. I saw him three days later at the aid station at Three Tango. His only problem was trying to get rid of the ringing in his head.

The last week of March we were given the duty of guarding five 105mm guns that were about four miles from Three Tango. Their fire base was being moved and these guns had not been able to leave that day. It had been a large base. Our platoon was to guard the entire perimeter, and even though it had concertina wire and fortified bunkers around it we were only able to put two guys in every third position. It was far from ideal, and no one felt good about it.

We were told no one was to sleep that night. Our command track was positioned in the middle of the perimeter and I was going to pull radio watch. Around 2200, Three Tango began receiving mortars and we received word that most every unit within a ten-mile radius was being hit. The artillery guns began firing white phosphorus rounds into the trees around us in an attempt to start fires to give us some light.

Around 0100, Lt. Campolo told me to climb into a bunker behind the track and get some sleep. This sounded good to me. I had been looking through the star-lite scope for the past hour, and it seemed quiet around us.

When morning came, we saw logs across the wire on the perimeter and noticed that all our claymore mines had been turned around facing us. Most of the trip flares had been disconnected, and by the tree line we found arrows, made from sticks, pointing directly at every APC. The most shocking was the sandal prints in the mud that we found behind our command track, and passing just in front of the bunker I was sleeping in.

It had probably been a squad of NVA who never had the capacity to attack with violence, but they did have the ability to attack psychologically. This was the most brazen act by the enemy we had seen. We could not visualize being that bold. It clearly indicated a difference between us and our enemy.

I may not be walking and sleeping in the jungle any longer, but the danger was still around me. Dr. Sallel had told me I could come back to the aid station if I wanted to. Our replacements were there, and he felt those of us who had come over with the original group could begin doing the easy jobs. Being on our own appealed to me more than my common sense.

We were hitting mines quite regularly now. So far, no one had been injured seriously. Our daily routine was consistent, and we began each day at 0700 by clearing the road. One day we were late getting started and it was close to 0900 when we left the fire support base. We were about halfway between the fire base and Three Tango when we noticed newly dug positions on both sides of the road.

We had just started down a straight stretch of road that had a bank on our right side, which was about twelve feet high, and the left side went downhill into the jungle. The guys got out of the tracks and began checking each position. Most of them had gone to the right side.

Lt. Campolo was sitting behind the 50-cal machine gun and had been watching those on our left. He climbed out of the turret and told me to take his place on the machine gun. Without thinking about it I climbed up and swung the gun towards the left side. I watched as he walked over to the tree line and headed down a trail. It was then I noticed his .45, CAR-15, grenades, and even his helmet lying beside the turret. He had left the track without his weapon.

I told the radio operator to get up on the gun and I jumped off the track with his weapon, several grenades, and a couple bandoliers of ammo, and headed down the trail after him. I found him about two hundred feet into the jungle. When he saw me coming, he headed on down the trail. It wasn't long before we were standing beside a stream. The road was about a quarter of a mile up hill, and in the stream bed were dozens of sandal prints. He told me to stay there and wait for him.

As he crossed the stream, I crouched into the fold of a giant teak tree. Tucked in against the smooth bark, hidden in the fold of the tree, I couldn't be seen unless someone had been watching me. Now there was nothing to do but hide in the humid jungle and wait. I was now completely alone, and I still had the Lieutenant's weapons.

The jungle around me came alive. Monkeys screamed in the tops of the trees, and a lizard scurried up the tree across the trail. I looked into the thick vegetation trying to see. I listened to the jungle noises trying to hear. Trying to detect any change which might indicate that the enemy was near.

As I listened, I realized once again that Lt. Campolo was not armed. I was the one who had the weapons. It was then I heard that still small voice inside asking, "Ron, what are you doing?" I had put myself into a position which went contrary to the commitment I had made with God. A situation which could cause me to take another's life. God left me there for about ten minutes to think about what I had done.

Suddenly, the mood of the jungle changed. Coming down the trail I heard pounding footsteps. I took aim at the bend in the trail, and before my heart had time to stop I saw Lt. Campolo slide around the corner and leap across the stream. As he passed me he said, "Let's get the hell out of here." I bounded out of my hiding place and took off to catch up with him.

We had not gone far when my legs began to feel like rubber. We were running up hill, and I was continually checking the trail behind me to see if we were being followed. All the while expecting to hear weapons open up behind us. When we arrived at the road, we half collapsed, sweat dripping from our face, and our mouth feeling like cotton. Lt. Campolo spread out his map and breathlessly called in artillery and air strikes. He had found the NVA's Base Camp. Luckily, he had not been seen.

I began to think about what had happened. I did not want to be put in a situation to have to kill anyone. I realized that what I had been through over the past eight months, and despite my prayers to be protected, if it came down to kill or be killed, I would kill. And that thought bothered me.

Working out of the fire support base had given me the opportunity to see my medic friends on a regular basis. The line companies had been spending two weeks on patrol and one week at the fire base. By the third week in March, most of the 2nd of the 8th had been taken out of the field and were back at Base Camp getting their new tracks. At the end of March, Dr. Sallel assigned Dio Rader to recon along with me. We were getting so active with convoy duty that our tracks were being constantly split up.

During the first week of April, I stopped at the aid station at Three Tango and heard that Danny Patton had been wounded. He was one of our original group and had been assigned to Charlie Company 1st Battalion 22nd Infantry when we arrived in Vietnam. It had been a few months since I had seen him. Whenever he was on patrol he always walked in the same position. One afternoon the guy walking behind him asked if he wanted to trade places? He said he was getting tired of the scenery, so Danny had traded with him. Twenty minutes later an artillery round landed beside the guy, killing him instantly, and tearing Danny's leg up pretty bad.

As I rode back to the fire base, I began thinking of our original group and who all had been wounded or gotten sick and was no longer in Vietnam. I realized that I was one of the more fortunate ones. A week later we had another track hit a mine. This time it was almost dark. Behind the track was a hill with a line of trucks loaded with 105mm shells. There was a Captain riding with the convoy and he had given the order that all trucks were to have their wheels blocked and the brakes on, and the drivers were not to get out of the trucks.

We had been working for fifteen minutes trying to get the road wheel off, so we could pull the wounded track back to camp. Someone looked up just in time to see a truck load of 105mm ammo rolling towards us. We watched as it began veering to the right, and just before it hit the track, it went over an embankment and rolled on its side.

I don't think I had ever seen an officer get so angry. Livid would probably describe it best. I wondered why the guy standing beside me was so nervous, and I soon found out when the Captain calmed down long enough and asked to see the driver. I thought he was going to kill the poor guy on the spot. The driver recovered the radio and his weapons as we formed a line and began hauling ammo up and loading it on another truck.

Lt. Campolo handed me his CAR-15 and told me to go over by the tree line and keep my eyes open. Without even a pause I took it and headed towards a teak tree. I tucked myself in its folds and began peering into the dusk. And once again I heard that voice inside of me. I couldn't believe I had done it again.

We finally got the road wheel off and began limping towards the fire base. A platoon from Charlie Company came to pull guard on the truck and what ammo was left. I never did hear what happened to the driver of the truck, but I'm sure the last part of his tour wasn't as pleasant as the first part.

When morning came, I had made my decision. It was time for me to take Dr. Sallel up on his offer. Around 1400 we were sent to Three Tango to get a convoy that was forming. When we arrived, I told Lt. Campolo what my plans were, and he told me he was going to miss me. There were things I would miss, but twice I had gone contrary to my convictions, and that was important to me.

I carried my pack into the aid station. Dr. Sallel looked up from a patient he was checking and asked me if I had had enough. I said yes, and he turned to a new medic and told him to get his things. I walked my replacement over to the command track and introduced him to Lt. Campolo. After seeing the convoy off I felt a weight lift from me. I knew I could still end up out in the field before my tour was over, but at least now I was doing what I felt was best.

When I got back to the aid station, I found Corky was there as well. He had come in from the field a couple of weeks earlier. He told me the unit that replaced us when we had gone mechanized had made contact last night, and while the medevac was making a night pick-up a parachute flare had gotten tangled in its rotors. It had been hovering about twenty feet, and pulling a wounded aboard, when it crashed. The pilot, the wounded guy, and the medic had been killed. So far there was a reported fifty-three WIA's, and twenty KIA's.

My arrival at Three Tango came just in time to move. The next day I was sent with the advanced party to establish a new base at The Oasis. This was a good move being as it was in more open country and was a lot closer to Base Camp in Pleiku.

105mm artillery firing white phosphorus at tree line in attempt to ignite fires

Chapter Nineteen

"I've entered the snapdragon part of my life. Part of me has snapped and the other part is draggin."
—Laurie Denski-Snyman

We were soon settled at our new base at The Oasis. The first three weeks were spent filling and stacking sand bags, and after the new bunkers were built, things leveled off to a routine of sick calls, and medevacs. My primary duty during sick call was giving penicillin shots for VD and drawing blood from malaria patients.

We were only a half block away from 4th Med, and whenever a medevac would come in Dr. Sallel and I would go and assist. It was assisting with these medevacs that I enjoyed most. 4th Med was the closest facility to the combat areas, and in a lot of cases we were the first stop once the wounded left the battlefield. We would do what we could to stabilize the guys, and then evacuate them to the 18th Surgical Hospital in Pleiku.

One afternoon Dr. Sallel and I were working on a guy who had taken an AK-47 round in his abdomen. The slug had entered the upper body and exited through his collarbone. The patient was slipping in and out of consciousness. Dr. Sallel made a small incision between the ribs and was in the process of forcing a small plastic tube into the lung cavity, when another stretcher was brought in and put on the floor beside me.

I looked down and into the eyes of an NVA soldier. He was bathed in sweat and was looking up at me. Both of his hands were cupped at his abdomen to keep his intestines from falling out. I turned my attention back to the one we were working on, and a moment later I looked back at the wounded enemy, and he was dead.

A lot of my friends from Alpha Company were taking their R&R and would spend a night at The Oasis on their way back to camp. One night I was listening to them talk and was thinking about how we had all changed since that first day we arrived in country.

The discussion was getting a little heated as I watched them drinking warm beer and arguing about the fine art of killing. I heard one of them say that killing an enemy soldier, no matter if it was a woman or a man, was part of war. If you were going to fight a war, then people were going to be killed, and if you were not prepared to kill, then you had no business being there.

As I walked away I thought, "How true". This war was not like those I had read about in history books. We were not an army marching through the country conquering as we went. We were tired, dirty young men trying to stay alive for just another few months, and then go back home where we could try to forget what we were experiencing.

During April, we began seeing weather changes which indicated the monsoon season was not far away. We were told we could expect twenty-four rainy days during the month of May, and June and July would just be rain. The dust around The Oasis was six to eight inches deep in places, so some rain would be welcome.

I was in Base Camp getting supplies for our aid station when a driver from transportation walked in and asked if I was the one who had ordered a new Corvette from Universal Auto Sales. He said he had ordered a new Pontiac, but a friend of his, who had gone home a month earlier, had written and told him the dealer had never heard of the order, and his car had not been delivered.

This was all I needed to hear. Before I went to sleep I wrote a letter to the dealer in Seattle, and the Chamber of Commerce in San Francisco. The letter to the Chamber of Commerce was asking them to check the stateside address. I had just mailed my $60.00 payment and I wished now I still had it.

The next day I got an appointment with the Adjutant General's office to ask an attorney what he thought I should do. Once I explained my problem I knew I was in trouble. I think it was the smile on his face as he reached in his desk and brought out a thick file folder that gave me my first feeling that things were not good.

He explained that the army had begun an investigation on Universal Auto Sales, and they were finding that very few cars were being delivered. The best thing for me to do was to stop making payments. He drafted a letter for me to send to the Hong Kong office, and asked that I give him a copy of any correspondence I might receive from the letters I had sent. As I walked back to the aid station I thought of many things I could have used that $340.00 for.

Since my return from the field I had been able to concentrate on my letter writing, and it was beginning to pay dividends. My short timers' calendar was now under ninety day and I was beginning to formulate plans for my return to civilization.

The first week in May, Alpha Company made contact, and before it was over the enemy body count was close to two hundred. They had counted over twenty officers killed, and among those was a Chinese officer wearing a Russian made watch and carrying a Russian pistol. Corky and I listened to the action on the radio, and waited for the medevacs to come in. These were our old platoons and our friends. When the battle was over they had three KIA's, and only eight WIA's, and these eight were only wounded slightly.

A few days later, Dr. Sallel and I walked in the front door of 4th Med just as a guy collapsed. He had been receiving a series of penicillin shots for VD and was going into anaphylactic shock. We picked him up and laid him on the stretcher. He was conscious, but having difficulty breathing.

He was given a shot to counteract the penicillin, but before it could take affect the guy got to where he couldn't breathe. He began fighting to get up and his face was turning blue. It took six medics to hold him down while I poured a disinfectant over his throat, and Dr. Sallel performed a tracheotomy. I had seen this done on training films at Ft. Sam but seeing it for real was quite the experience.

The next morning, I was drawing some blood when in walked Harold. He had just returned from Trippler Army Hospital in Hawaii and was heading out to my old platoon. I could tell recon was not his first choice. He had spent two months in Hawaii with malaria and had just returned from a couple of weeks leave at home. Southeast Asia was the last place he wanted to be.

Clarence Lipscomb was with weapons platoon in Charlie Company when he left for his R&R in Hawaii. Charlie Company was back at The Oasis getting the last of their tracks, so I was sent to take his platoon while he was gone. I was hoping he would get back before we headed out to the field again.

On the 20th, the entire company, except weapons platoon,

was sent to reinforce a South Vietnamese army unit a few miles to the west. We were told to standby in case we were needed. On the 21st, Corky and I listened to a battle taking place between Alpha Company from the 1st of the 8th. They were still on foot and were patrolling the border around the Ia Drang Valley.

The company had made a combat assault and had landed at what they thought was a cold LZ. Weapons platoon was left to guard the LZ and the rest of the company began patrols around them. The NVA waited until they were away from weapons platoon and then pinned each patrol down. Once this was accomplished they overran the LZ.

The last radio contact with weapons platoon was someone saying that everyone else appeared to be dead and requested that the artillery saturate the LZ. The next day when Alpha Company was finally able to break contact, they had suffered forty-two KIA's and eighty-three WIA's.

Clarence returned to The Oasis on the 23rd, and I breathed a sigh of relief as I put my gear back in my tent. I had been with his platoon for two weeks and we had never left The Oasis.

Harold was with recon on a mission moving Montagnard's from their jungle villages to a new resettlement village closer to Pleiku. One evening he got word that a Montagnard family was at the perimeter and wanted to see a doctor. When he got there, he found an older man that looked like he had been pulled through a buzz saw. He had been attacked by a tiger.

Harold did what he could then called for a medevac. When it arrived, he had the whole family of six get into the chopper. There happened to be a new Lieutenant who didn't agree with Harold's decision, and decided to write him up on report for the misuse of government equipment.

The next day they both were advised to fly to Base Camp to see the Lt. Colonel. When they were seated in his office the Lt. Colonel looked at Harold and winked. He then turned to the Lieutenant and proceeded to tell him that there was two people in his command he was not to mess with. One was him, and the other was his medics.

To Harold's surprise, he found that the Intel people had been trying unsuccessfully to get intelligence information from this village for weeks. The old man was a leader of the village, and since Harold had helped him he was giving information that would save a lot of lives for both his people as well as us.

A week later, recon was patrolling close to this village when they noticed a large tiger stalking them only a hundred feet away. They opened up on it with their M-16's and, to their surprise, the tiger attacked. They shot its foot off with the 50-cal machine gun and it still came on. Someone finally stopped it with a .45 just as it was getting ready to leap onto one of the tracks. It was believed to be the same tiger that had attacked the old man. Prior to the old man, it had killed three other people from the village.

Soon after we set up camp at The Oasis, the engineers had begun building an airstrip large enough to handle C-130's. On the morning of the 26th, they began landing, and throughout the day units of the 173rd Airborne arrived. By this time enough of the 2nd Battalion 8th Infantry was getting their APCs that our AO was being taken over by other units. They were setting up a forward Base Camp about eight miles east of The Oasis.

Corky and I had just returned to the aid station from watching the planes landing when we got a call to return to the strip. One of the Airborne guys had walked into the prop of a C-130. Dr. Sallel took off in the jeep and we headed across the camp on foot.

As we ran we had visions of finding pieces of someone. When we got to the plane we found him lying under the propeller, which by this time had come to a stop. He was in one piece but had an inch and a half groove taken out of his forehead from just above his left eye and running all the way to his left ear. Amazingly he was not only alive, but conscience as well and just wanted to go back with his platoon.

Dr. Sallel arrived just as we got there, so we loaded him onto a stretcher and headed for 4th Med. While the doctors did what they could for the guy I talked to him and got the information needed for the medevac tag. He was deathly afraid of getting a shot, and only wanted to return to his unit. As far as he was concerned he had only been knocked on the head by something. He probably would have died of shock if he could have seen himself in a mirror.

When you exit from a C-130, you leave from the rear of the plane. The guys were leaving the plane and walking away from it to waiting trucks. It seems this guy was air sick from the flight, and for some reason he had turned towards the front of the plane and had walked into the prop as it was coming to a stop. I was amazed that he had lived because the skull was jagged and completely gone in places. He was one lucky trooper.

We had no sooner finished with this guy when another medevac landed. This guy had been walking on patrol when he accidently pulled the trigger on his M-79 grenade launcher. The safety was off, and he had fired a round into the calf of his right leg. Luckily for him, the grenade will only arm itself after it rotates so many times. In the short distance between the weapon and his leg it had not had time to do this. The round was firmly lodged in his leg. Dr. Sallel took one look at this one and decided to let 4th Med deal with it.

The prediction of a rainy June became a reality, and along with it was an unusual amount of malaria patients. 4th Meds malaria ward was filled with tubs full of ice floating in alcohol water, and in each tub was someone shaking uncontrollably, and submerged up to his chin. And if this wasn't torture enough, large fans stood at the foot of each tub blowing cool air. Even with this treatment it was hard to get some of the high temperatures down.

On the 15th, Harold once again tested positive and headed for the malaria center at Cam Ranh Bay. The sad part was the two to three months of recuperation was coming at a point when we were under thirty days on our short timer's calendar. Even sadder was the day after he left for Cam Ranh Bay his rotation date was changed to June 23rd.

Rotation schedules were changing daily. Some of those who were scheduled to leave the end of July were leaving now to be reassigned in the States. If we were under ninety days when we returned home, we would be getting an early out. It appeared my discharge date was going to keep me in country for my full time. I didn't know what was worse, having to stay my full time or having to be reassigned and staying in the Army until the middle of October.

A lot of my time was being spent helping Dr. Sallel with the rotation physicals. Shot records had to be current, and blood draws were done on everyone. When I was at Ft. Sam I had thought it would be fun giving shots and drawing blood, but now it was getting old.

Just after lunch on the 27th, I was sitting in the aid station talking with another medic when in walks the sergeant from headquarters. He looked around then asked if we wanted to go on another R&R. This was too good to be true. He had two slots to fill for Taipei, and had just a few minutes to do it in. We felt it only polite to accept the offer. He then told us we had twenty minutes to catch the chopper.

That evening we were at the R & R center in Cam Ranh Bay, and on June 30th we boarded a Pan Am prop-jet and took off. Once we were airborne we were given an iced hand towel and a bowl of tropical fruit. A few hours later we landed in Taipei.

Taipei was small compared to Hong Kong, but the sights were similar. I spent my time walking the streets and looking in the shops. The Navy compound was large and filled with things for us to do. My five days went by fast and when it was over I was ready to leave. I knew that when I got back to The Oasis it would be almost time for me to fly home.

Chapter Twenty

"We live in a world that has walls and those walls need to be guarded by men with guns."
— Aaron Sorkin, *A Few Good Men*

When I returned to The Oasis, I found my rotation date had been moved up, and I had missed a freedom flight. This was a disappointment, but I had seen another country, so the tradeoff was acceptable. A freedom flight was a commercial flight set up for those who lived in the northwest. My next opportunity was going to be on a C-141 military transport sometime in the next couple of weeks.

Most of our replacements that arrived in June had gone to replace those lost when the 1st of the 8th had been overrun. We were now getting our new medics almost daily, and with so many people just hanging around we did our best to live up to our motto of "Out of sight, out of mind."

The next two weeks went by fast and on the 16th I was told to get my gear together and head for Base Camp. I stayed at the aid station only one day before I moved on to the rotation center on the northeast side of Pleiku. As we drove through Pleiku City, I realized that I was looking at it for the last time.

Our flight was scheduled to leave the morning of the 19th, but the monsoons had other plans for that day, so it was rescheduled for the afternoon of the 20th. Finally, on the morning of the 21st, we were standing on the runway as a giant C-141 tried to land. The monsoon clouds were closing in fast and our ride was having difficulty getting down. We were told it was going to try one more time. If that failed, we would have to be flown to Cam Ranh Bay by C-130.

We watched as the plane came around for the last time, and to our relief it touched down. Within an hour, we had landed in Cam Ranh Bay to top off our fuel tanks before heading for Japan. As the wheels left the runway, a cheer echoed through the plane. We had spent our year in combat and won. Our war was over.

Hours later, we were landing at Yokota Air Base in Japan. The snack bar was getting ready to close as we walked into the terminal, so we made a dash for the door and convinced the manager that they should stay open a little bit longer. Before the hour was over every hamburger, milkshake, and fries in the place had been purchased.

When we made our way back to the plane, we were told we would have to spend the night in the hanger while our plane was being repaired. It seems while we were landing we had lost some part that had to do with the oil pressure. We were put in a hanger that was being used for storage. It had no chairs or benches, and by the time morning came we were ready to do anything to get on our way. Around 0400 the decision was made to prepare another plane to take us the rest of the way, and by 0730 we were airborne once more.

It was a long thirteen-hour flight. We could feel the plane losing altitude and I knew we were getting close. I stood on the seat and looked out the window. Below me was a forest filled with fir trees. Something I had not seen in a year.

We circled McCord Air Force Base and came in for the landing. I could not stop the tear that trickled down my face as I walked off the plane. I looked around to see if anyone had noticed, and saw others trying to hide the same thing.

As we walked into the terminal I saw my family standing in the crowd. What a beautiful sight to see. I remembered a year earlier when I had stood at the railing of the troop ship as we passed Seattle and wondered if I would ever see that beautiful city, or my family again, and now here I was, home at last.

I was given permission to go home and return to Ft. Lewis the next day for processing out of the Army. Dio Rader was on the same flight, so he came home with us. My sister had brought my car, so I could drive it home. I had not driven much while I was gone and when I did, it had been a jeep. I killed the engine five times just getting out of the parking lot.

Dio and I were at Ft. Lewis by 0800 the next morning ready to process out. By the time we were finishing, we had caught up with a lot of those we had flown with the day before. They had been waiting in lines most of the night.

My tour of duty had, without a doubt, been an exciting one. I was given the opportunity to see parts of the world I would never have seen otherwise, and I had the privilege of meeting friends which would stay friends for a lifetime. I have read somewhere that there is no greater bond between men, than the bond shared by those who have faced death in battle. I'm not sure how correct a statement that really is, but I do know there is a bond that is forever.

There are many who experienced Vietnam, and still linger in

its shadow. Many returned wounded in mind, as well as body, still others were unmarked physically, yet carry the scars of emotional pain. I watched, along with the whole world, the fall of Vietnam, and later the returning POW's, and I watched the nightly news showing flag draped coffins being unloaded from giant military aircraft. In the fear and gore of war, adolescent men are tempered and made hardened. Those of us who experienced Vietnam came home a different person than when we left.

In Washington, D.C. there is a black granite wall, and on its face are engraved the names of everyone who gave their life during the Vietnam War. At most anytime you can see someone meditating there, as if in a church. Their head bowed as they wipe away a tear, or their finger lightly touching a name. Their thoughts many miles away. Thoughts that span continents, and time. Thoughts of a moment long ago when a portion of their life died along with that special name on the Wall.

The names represented there each have a story to tell. Stories of violence, and pain. With each death, many lives were changed. Parents, spouses, lovers, children, neighbors, the list is endless. It doesn't matter who the person is, they will never be the same again. Fifty-one years later, just the mention of the Vietnam Memorial, or remembering Allen's death, still will bring a tear to my eye.

Lives lost, for reasons which are now hard to understand. There were some who died within hours after reaching Vietnam, and some who died just hours before leaving. The Vietnam Memorial shows the names of those who died, but what about those thousands who will spend the remainder of their lives scarred physically and emotionally.

In our lifetime, we have known many walls. If you were raised a Christian, your earliest recollection would probably be the walls of Jericho. There is the Great Wall of China, and the Wailing Wall in Jerusalem. In more recent times there was the Berlin Wall. But none will ever have the impact on the twenty first century America as the Vietnam Memorial Wall.

Epilogue

Many years have passed since the late afternoon in 1967 when I walked into the McCord Air Force Base waiting room and saw my family waiting for me. I never had to experience the hostility many returning soldiers went through. I was never called baby killer or spit on, but I watched on television as this happened to others.

Both my Mom and Dad have passed away now and my sister and I are senior citizens. My daughters are adults and I have grandkids in their twenties. One would think my Vietnam experience would be a fading memory, but I can honestly say that I have thought of Vietnam every day since my return. The nightmares ended many years ago and with the VA's help I now get a good night's sleep.

When I graduated high school and watched my friends leave for college, I made the decision that someday I too would go to college. After a successful career in sales of aluminum in the aerospace industry, I did begin college in the summer quarter of 1988. What a wonderful time that was. I was a mature adult and college was like a job to me. When I completed my college experience, I held a Master of Social Work degree and began a second career in mental health therapy.

My last job was as a therapist with the VA helping Vietnam

Vets who were diagnosed with combat related posttraumatic stress disorder. This was the most rewarding job I ever had. Like many of you, I had been reading books, like this book, of guys just like me. I had just finished reading about the battle at Dak To in 1968 when I had a new Vet come to my office. He had been on the perimeter guarding the command group and medics caring for the many wounded when the jet accidently dropped the five-hundred-pound bomb on them.

With tears in his eyes he said, "And there was this Chaplin who always went with us on patrols. He was with the wounded when the bomb hit and was killed. I've spent the last thirty-three years trying to remember his name." I said, "Chaplin Waters." He looked at me with astonishment and said, "How did you know that." I told him I had read a book that told about his experience. I was deeply saddened when my own PTSD hit me, and I had to medically retire.

I have stayed in contact with several of the medics I trained with. Sadly, Harold Stenseth and I attended the memorial service for Frank "Corky" Colburn after his death from cancer, caused by agent orange. Each of us have been diagnosed with PTSD and either have cancer or type ll diabetes, also caused by agent orange.

Every October I've thought of my first combat. I've wished I had the name of the guy who ran back outside the perimeter and helped me carry the wounded guy back to safety. In October 2017, I decided to try to locate him using Facebook, so I posted the following on a dozen Vietnam Veteran websites.

October 25, 2017—National 4th Infantry Division Association Group.

There has been something on my mind now for several months and I know this request is a long shot, but I would like to ask anyway. When the 4th Infantry Division went to Vietnam I traveled on the USNS Pope.

I was a medic attached to Alpha Company 2/8th Infantry. My first combat was towards the end of October 1966 when I was the medic that came with my platoon Sergeant, the LT and seven others to reinforce Bravo Company who was hit and surrounded on a small hill. The first night I was there we were hit around 9:00 pm. I was in a foxhole with the Sergeant and LT half way up the hill when my first call for a medic came just moments after hell broke loose. It came from down in front of us.

I jumped out of the hole and crawled down the hill to the perimeter and asked who called for a medic and the guy pointed out in front of us to a body laying very close to the jungle. The guy had been one of three who were on a listening post. One guy made it inside the perimeter, the one guy out in front of us and the other guy died at the LP.

I ran out to the guy and knelt beside him. He was laying on his stomach and I could see in the moonlight a blood stain on his hip. I ask him if he was hit anywhere else and he said, "Get down they're right behind us." I gave him a shot of morphine and ran back to the perimeter and said that I needed help and a guy from Bravo Company jumped up and we ran out and carried the wounded back inside the perimeter. It is a miracle we were not KIA.

I am a Christian and was one of the medics who never carried a weapon. Those of you who were in Bravo Company will remember that most of the medics in your company also never carried a weapon. I can only remember several names of your medics, Harold Stenseth, Dio Rader, and Billy Winters.

You also had a Sgt. Turner who had served in WWII and

Korea and was awarded a Silver Star for helping one of your LT's rescue several guys that were pinned down outside the perimeter on the first night. Your CO called him "Papa T" and after this battle he was reassigned to a safe job at The Oasis taking care of the Sun Bear Cub "Bravo".

Here is my question. I never thought to get the names of the wounded guy or the guy who went back outside the perimeter with me to carry the guy back to safety. If any of you who were on that hill and might remember this, I would love to talk with you.

On November 10, 2017, I posted the following:

Six weeks ago, I began a journey to see if I could locate two members of B-Co. 2/8th Inf. 4th Div. who had shared an experience with me during my first call for a medic, which occurred on the night of Oct. 29, 1966. I never knew their names but took a chance to see if I could locate them.

And it worked. Through the assistance of Richard Bolin, who was on the hill that night, and his wife Sheila, I was able to visit by phone with Preston Leaderbrand and his wife Bonnie. Preston was the one who jumped up and ran back outside the perimeter and helped me carry the wounded guy back to safety.

Preston told me he couldn't remember the wounded guy's name but had heard he did pass away a couple of days after being medevaced off the hill the next morning. Preston was one of two RTO's for B-Co. CO and wasn't needed so was assigned to assist the medics if needed. When I ran back to the perimeter to get help I never knew I was being guided to the position that had someone who was willing to risk his life to help me.

What an amazing journey this past six weeks has been. I would like to thank each of you who made such nice comments on my post and encouraged me.

But the story doesn't end there. While visiting with Preston and Bonnie that evening I learned that they attend the yearly 4th Infantry Division Reunion and were planning to attend the 2018 reunion in Green Bay, WI in late July 2018. After talking it over with my wife Bev we decided it was time for a road trip. We packed our things and headed for Green Bay. What a great time we had meeting Preston and Bonnie. Also, in attendance was Richard Bolin and his wife Sheila, and Roger Dearmyer and his wife Sue. Both Richard and Roger were also on the hill that night.

Pretty exciting after 50 years. So, as the story began, it will end.

"Where shall I begin," he asked, "Begin at the beginning," the King said, "and stop when you get to the end."
—Lewis Carroll, *Alice in Wonderland*

The End

About the Author

Ron Donahey was born in 1945 in New Mexico. At the age of five, he moved with his family to Southeast Alaska, and at age ten, they moved to Washington state. Ron was drafted into the Army and served in Vietnam as a combat medic with the 4th Division. Three days after turning 21, he was in combat for the first time. At the age of 43, after a rewarding carrier in the business arena, he went to college. When completed, he held a Master of Social Work degree and began a second career in mental health therapy. He specialized in the treatment of veterans diagnosed with combat related Posttraumatic Stress Disorder. He retired in 2004. In 1990, he wrote of his army experience and his manuscript sat on a shelf for the next 28-years, until he was encouraged to submit it to a publisher for review.

www.ingramcontent.com/pod-product-compliance
Lightning Source LLC
Chambersburg PA
CBHW030111100526
44591CB00009B/366